New Times, New Rules:
Take Control of Your Farm Marketing

Scott Stewart

ISBN 1-58874-576-7

Copyright © Scott Stewart 2006

Published by

Stipes Publishing L.L.C.
204 W. University Ave.
Champaign, Illinois 61820

Stewart-Peterson
137 South Main St.
West Bend, WI 53095
(800) 334-9779
www.stewart-peterson.com
email: info@stewart-peterson.com

TABLE OF CONTENTS

Chapter ....**Page**

Table Of Contents i

Introduction 1

About This Book 3

Why Should I Read This Book? 7

Section 1: How To Successfully Market Your Production 15

1: Why Is Farm Marketing
   So Difficult? 17

2: A Known Fundamental Is A
   Useless Fundamental 23

3: Why You Can't Predict Prices 27

4: Strategy Is The Key 31

5: Manage Opportunities
   As Well As Risks 39

6: Our Basic Strategy 43

7: What Prices Do
   Is All That Matters 49

8: Always Be Diligent 53

9: Dangerous Yet Useful Tools 55

10: Farm Sizes Are Up –
    But Where Are The Profits?...... 59

11: Farm Marketing Has Changed.... 63

12: What Does It Take
    To Market Well? 71

13: The Strategic Approach
    To Marketing: An Outline 79

**Section 2: Market Concepts
You Can Use 93**

14: One Factor Drives The Market ... 95

15: Big Supplies, Small Supplies –
    What Do They Mean? 99

16: How Important Is
    Knowing Your Breakeven? 105

17: Price Level: Unimportant! 111

18: How to Use The News 113

19: Take Out The Emotions And
    Get It Done 115

20: Bankers: You Can't Blame Them! 119

21: Can You Trust Your Elevator? ... 123

22: Don't Try To Outguess
    The Report.................... 129

23: Don't Be A Sheep 133

24: Open Your Time Window 137

25: Carryover Stocks:
    Market Shock Absorbers 139

26: Buy Strength, Sell Weakness 141

27: Marketing: An Input Expense ... 145

28: Options Always Lose Money,
    So Why Use Options? 151

29: How About Using
    Technical Analysis? 157

30: Inflation And Other
    Insignificant News 161

31: What About The Strength
 Of The Dollar?................ 165

32: Bell Curve Your Sales 169

33: How Powerful Are Those
 Big Funds?.................... 173

34: Success Breeds Success:
 Avoid The Killer Loss 177

**Section 3: Where Can You Turn
For Help?......................179**

35: Marketing Advice Pays 181

36: Do You Want Professional Help? . 185

37: Use The Right Tool............ 189

38: Being A Believer 193

39: Setting Goals And Being Focused . 197

Appendix A: Glossary 199

Appendix B: About The Author 209

INTRODUCTION

For the first 20 years that I was in the business of advising farmers on how to do commodity marketing, I saw very little change in the markets overall. While I was being told about big changes taking place, little surprised me during that time. Prices had their typical up and down swings, but maintained pretty predictable price ranges. While markets were volatile and price ranges seemed wide, that was nothing in comparison to what we have seen since then.

The really big changes came with the bull market crops of 1995/1996. Since that bull market, the world of commodity marketing has changed dramatically. Brazil has moved from being a supplemental producer of soybeans to a dominant competitor. China's movement in and out of the marketplace has had significant influence. Crop yields and overall production potential have gone through the roof, while at the same time, demand has increased proportionately. The end result is that, for most ag commodities, price ranges from high to low have doubled or even tripled.

As a result, the use of futures has a greater potential and a greater risk than ever before. Our markets have become global, with a resulting increase in volatility. Another consequence is that option costs have increased dramatically.

Because of all these dramatic changes, your marketing methods and approach must also change dramatically. The pain and loss of poorly executed marketing, as well as the rewards accrued from a professional marketing approach, are greater now than ever before.

ABOUT THIS BOOK

The objective of this book is to give you a new perspective on your approach to commodity marketing. I want to dispel some old myths, help you understand why marketing is so difficult, and offer some valuable tips on easing that difficulty. This book will introduce you to the tools that are needed to become an efficient marketer in this new age of volatility. I want to encourage you to take a serious look at your marketing and evaluate how you are making your current marketing decisions, how you are implementing those decisions, and what tools and resources you are using to accomplish your marketing objectives. Then you will need to assess whether your current approach is doing all that it can for you.

A disclaimer is in order here. This book is not a general text to educate you on the ins and outs of futures, options or cash marketing tools. Rather, it will give you the framework and perspective to move forward in making sound, professional marketing decisions.

There is no attempt to hide opinions or argue all sides of every issue. The goal and purpose of this book is for me to try to convey to you how I believe you need to look at marketing to be successful in the future. I want to motivate you to take action and make the changes necessary to be a successful agricultural producer for many years ahead.

This is the same objective I have for my customers and clients. I want them to be the producers who survive and prosper in the years ahead, rather than those who are forced to sell out. Nearly every farmer has learned to be a very efficient producer. The successful producers of tomorrow will be the ones that conquer marketing, either by learning to do it themselves, or by gaining the knowledge needed to properly and confidently hire professionals to do it for them. No matter which path you choose, it will help assure your farm business of continued survival in the years ahead.

About This Book

In writing this book, I sat down with no predetermined objective as to how long the book should be, how many chapters it should be, or any other constraints that publishers typically put on an author. I simply listed out all the key concepts that I thought were important to convey to you. As you can tell by looking through the book, some chapters are only a couple pages and some are many pages. My only objective was to pass along to you, in the most clear and concise way, the marketing concepts that will be essential to your success in the future.

WHY SHOULD I READ THIS BOOK?

The Future of Agriculture

Oftentimes, when I am out giving speeches and seminars, people ask for my view on the future of agriculture in America. I'm including my answer in this book for two reasons: One, planning is easier with an idea of where this industry is going. Two, marketing will play a big part in your future of agriculture.

My outlook on agriculture? Agriculture will remain unchanged. By "unchanged," I mean "marginal." Average producers will break even, above-average producers will prosper, and below-average producers will fail. The number of farms will continue to decline substantially in the years ahead, just like they have declined substantially in past years.

If you look at USDA data for the past 30 years, farm income in the central Corn Belt is equal to USDA farm program expenditures. That means that the average farmer is only making what the USDA is paying them. Below-average producers are making even less than that. Not only are they not making money farming, but they are losing part of their farm program payments. Above-average farmers not only are making what the USDA is giving them, but they also are making a return on their farm investment. I do not see that changing.

Between the abilities of our U.S. producers, worldwide production capabilities and our technological advancement capabilities, I believe we will continue to produce the amount of food the U.S. and the world need, and there will be a balance. As food demand increases, prices will increase to allow for technological advancements to keep yields and production levels in line with consumption levels.

The story behind this outlook is that you have to be better than the average farmer to prosper in agriculture. In the old days, it was easy to be better than your neighbor – just plant and harvest a little more timely, manage your weed control a little better and maintain and adjust your planter and your combine. You had good odds of a pretty substantial advantage. Times have changed. Most farmers have good equipment. You can hardly find a field anywhere that is full of weeds. If it has a weed, you can spray it with a herbicide. The genetics of today's crops produce bigger yields more consistently than anything dreamed of 10 or 15 years ago. In short, the range of production capabilities between the best and the worst farmer in most counties is very, very narrow these days. What's more, it used to take weeks and months to plant a crop. With today's equipment, many producers can put a crop in the ground in a fraction of the time it used to take. With today's hybrids, in many parts of the country you can even harvest that crop without drying it, long before the snow flies. Agriculture truly has changed.

The Future of Independent Farming

Without getting too political, I think it is worthwhile holding up a caution flag and giving a bit of a warning about modern agriculture. I will put it this way: a vertically integrated, corporate American agriculture is going to be very bad for the individual farmer, and it will be very bad for rural communities. In the end, it will be very bad for the American economy.

What do I mean by a vertically integrated corporate agriculture? It is not a family farm that is incorporated, producing a crop or a specialty crop, processing and selling it. That is a good idea. Go for it – more power to you. What I'm concerned about is big agricultural corporations reaching down to the farm and offering financing, the supply of input needs and a whole host of other services in exchange for crop. This is already happening in South America, and it is a trend that is developing here in the U.S. It seems innocent enough. In exchange for agreeing to plant a certain amount of seed from one company and using their chemicals, you agree to give them a certain number of bushels. Oftentimes, these

Why Should I Read This Book? 11

programs are started on a limited number of bushels, like up to 20% of a producer's production. The attraction of guaranteed revenue, lower cash flow needs and all the other conveniences that go along with this are certainly an attractive lure to get a producer started down this path. Unfortunately, the only people that win in this game are the stockholders of the big corporation, who are sitting on their yachts or flying around in their private jets.

Once this vertical integration takes hold, the producer ends up being a worker bee for the big corporation. The producer commits all the crop, or at least such a substantial amount, that unless you have a windfall yield, you only make what the corporation has offered to pay you upfront. While that steady paycheck has some attraction, there are things to watch for. Ask any contract hog feeder how the fine print in those contracts came back to bite them. Ask those producers how the income compared over a period of many years.

More importantly, think about the impact on your own local community. When all your farm inputs are contracted through a large corporation and delivered directly to your farm, that revenue will no longer be flowing into your local community. When your tractors and combines are bought in mass and dropped off at your driveway, what impact is that going to have on the local dealership? When corporate America comes to the local farm community, all the dollars will be sucked out of the community and not put back into the community. The individuals that used to run the bank, elevator, machinery dealership, co-op, and all the other businesses that supported the farmers will disappear. Those individuals are often the individuals who had the money to donate for the hospitals, to donate for the churches, and overall, to be the community leaders that drive the community forward and make it prosper year after year. When all that money leaves the community, you're left with a lot of workers who, at best, can find a minimum wage job in a relocated factory. You, the farmer producer, will end up working for near minimum wage but still have a substantial investment and risk that is disproportionate to your income.

What I outlined above is certainly not a pretty picture. In a lot of ways, the vertical integration of our markets by large corporations may be inevitable.

In my view, the one way to at least postpone this trend is for all producers to do the most they can to maintain their ability to make individual marketing decisions and maintain their farm operation's profitability. If you are profitable, you will not need that helping hand from the big corporation. You will be independent and successful. The only way to do this is to either become knowledgeable and successful at strategic marketing, or hire professionals that can do it for you.

Section 1

HOW TO SUCCESSFULLY MARKET YOUR PRODUCTION

New Times, New Rules

Chapter 1
WHY IS FARM MARKETING SO DIFFICULT?

After being in this business for nearly 25 years and working one-on-one with producers, I have found a rather interesting pattern. Some of the very best producers do the poorest job of marketing. Why is this?

First, let's consider how the thought process for marketing works throughout a growing season. Figure 1 (page 18) shows a normal seasonal price pattern for the corn and soybean markets. Prices start out at the beginning of the calendar year with a slight rally into spring planting season and then fall into harvest.

At Stage ①, winter, there is a great deal of uncertainty, as nobody knows for sure how many acres will be planted. There are concerns that a spring frost could damage substantial acreage. There are fears that it is too wet, too cold, or too dry. There is uncertainty about exports; there is uncer-

18 New Times, New Rules

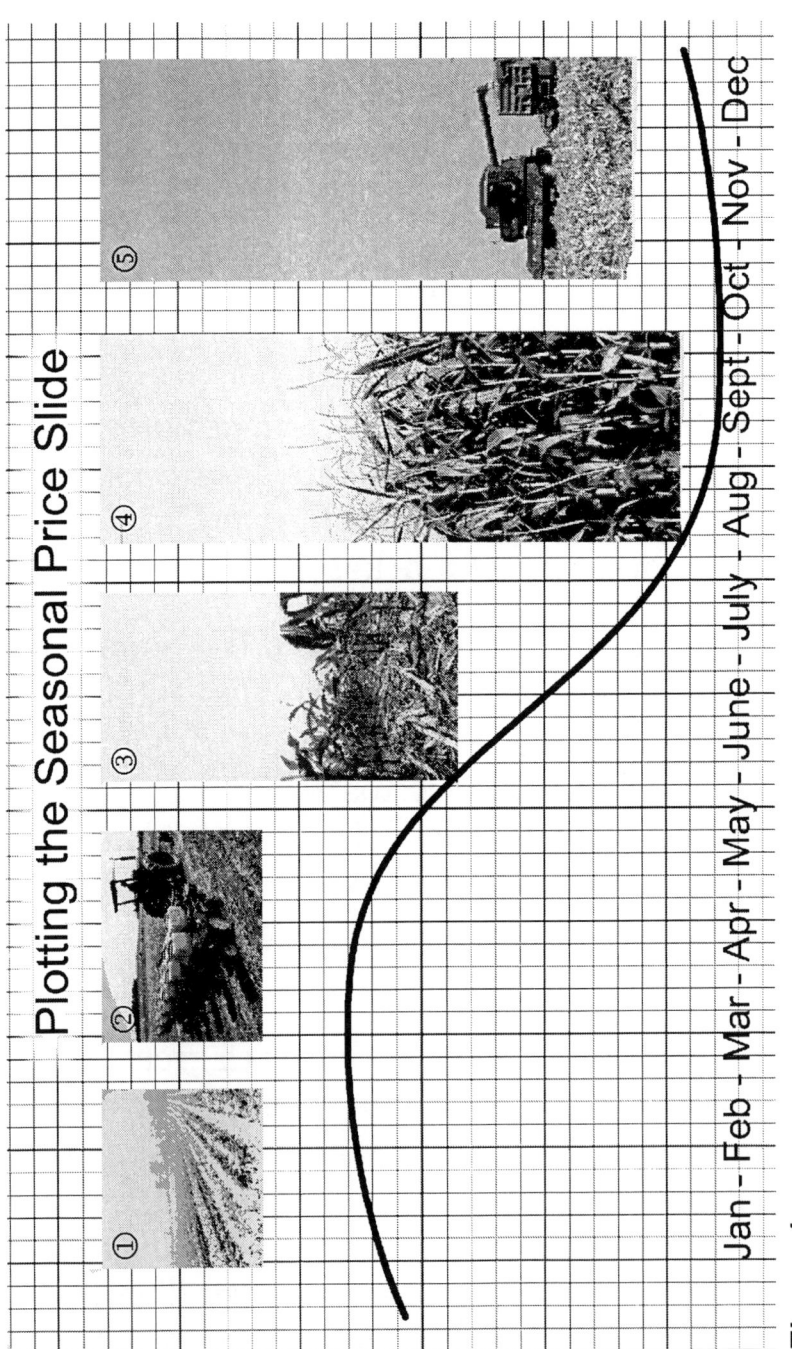

Figure 1.

Why Is Farm Marketing So Difficult?

tainty over whether there will be a spring frost, fall frost or whether rootworm will damage a lot of the crop; there are questions about China getting in or out of the market. The list goes on and on. These factors make it difficult to make a wise marketing decision. They make it difficult to know whether the price of corn is going to $5.00 and beans to $12.00, or whether prices are doomed to fall into a bottomless pit. This uncertainty is what makes prices rally.

Then what happens? At Stage ②, suddenly the weather breaks a bit, the tractors get into the fields and prices start to flatten out. We hear reports that the planters are rolling, and prices slip just a little lower. As the weeks progress, the seeds germinate and the crop begins to grow. We soon find that we have not had any substantial spring frost, and that the crop has broken through any heavy crust and is off and running. As the crop inches taller, the market inches lower. At Stage ③, the corn crop now is knee high, and prices have started a downtrend. By Stage ④, the corn has grown tall and healthy, production forecasts are rising, and prices are even lower. This is because there is less uncertainty in the market. We now know what acreage was planted, and we are starting to get yield predictions. Even the fears of a fall frost are starting to diminish, as the growing degree days mount up.

At Stage ⑤, you can see that the crop is being harvested and prices have bottomed out. We now know that there was not a fall frost; we know that we have had exports, and there was not a spring frost. We know what acreage was. We know, we know, we know. Unfortunately, what we really know all too well is that the best opportunity to price was many months ago, and the opportunity at hand is very unattractive.

Why are good farmers often bad marketers? Because good farmers usually are good businessmen. Good businessmen make good decisions based on good information. If they don't have the facts, they find the facts. If they can't get the facts today, they research the market and wait for the facts. They wait to make their marketing decisions until all the facts are in and they make sure they have good information.

Unfortunately, in farm commodity

marketing, by the time you have all the information you need to make a decision, the market move is over. Remember, a known fundamental is a worthless fundamental. (We'll talk about this more in the next chapter.)

So, everything that makes you a successful farmer, everything that makes you a strong businessman, everything that has put you in the position to be on the town board or the school board or the church board or the bank board has taught you to be an ineffective marketer. This is because when you wait for all the information to make decisions, you are waiting for the opportunity to disappear before you act. To be a good commodity marketer, you have to learn to make decisions when you do not have all the information. Let me repeat that: To be a good commodity marketer, you have to learn to make decisions when you do not have all the information.

How can you do this? Read on...

22 New Times, New Rules

Chapter 2
A KNOWN FUNDAMENTAL
IS A USELESS FUNDAMENTAL

In the spring of 2004, while I was waiting to give a presentation at Purdue University's annual crop workshop, I had the opportunity to hear a presentation by a fellow market advisor who used a phrase I will never forget. I do not know if he originated this phrase, but I have found it to be one of the more succinct ways of explaining what I have always believed. As the title of this chapter says, "A known fundamental is a useless fundamental."

Think about your own marketing experiences. When you have been absolutely convinced that prices have nowhere to go but higher, what do they typically do? Fall apart and collapse, right? Then, when prices have been declining for months and months and months, when you become completely discouraged and feel there is no hope that prices will ever go up again, and all the news is as negative as it could possibly be, you give up and clean out the bins, figuring that it is only going to get worse. Just about the time you haul the last load to the elevator, prices bottom out and start climbing higher. This pattern occurs time and time again, because when I talk about this at seminars, I see a lot of shaking heads and frustrated smiles.

What it all boils down to is this: when you can identify twenty reasons why prices should go up, you can just about bet that the price move is over. All of those reasons you can write down are known. The market has factored them in. They have been bid into prices, and the buyers have all bought because of those reasons. This is why prices have already risen substantially.

The other thing you should be aware of is what the media is doing when prices are rallying. When we get calls from the wire services asking "Why are soybeans up today?" we give them two or three reasons. They write it and publish it. You read it. It is all bullish. They do not call up and ask "Why should soybeans not have gone up today?", or "Why could soybeans fall tomorrow?", or "What reasons are there for soybeans to drop a dollar lower three weeks from now?" The reporters never think of that. All they are reporting on is today's market event. If prices are up, they want a reason for it.

Once all the bullish news is well known and is factored into prices, demand is choked off and prices start to fall. Then the reporter calls and asks, "Why are prices down?" We give him a couple of reasons. He is satisfied and writes his story. All you hear is negative news, and prices start to collapse.

A simple concept to remember is that we are talking about "futures" markets. Even tomorrow's cash price is a future market. Tomorrow's price expectation is based on today's news. Today's news is already factored into the market, so it takes new news to continue to drive the market. As they say, you have to continually feed a bull. When you run out of feed, the bull dies very quickly. That is why once a fundamental factor is known, it has already been fed to the market, it is built into today's price and it takes new news to drive the market to a new level.

Remember: A known fundamental is a worthless fundamental.

Chapter 3
WHY YOU CAN'T PREDICT PRICES

This entire book is based on one key concept. *Marketing should be strategic. It should not be based on a price outlook.* The whole foundation behind my approach to marketing is based on being strategic as opposed to being outlook-based. It is my underlying belief that you cannot predict prices. Sure, you can try, and it is definitely kind of fun, but in the end, predicting prices costs you money. Predicting prices takes your focus away from managing risks and opportunities.

Let's look at the crop markets and think about why you can't predict crop prices. Look back over 20 or 30 years worth of price charts. With only a couple of odd exceptions, almost every major high or low in prices can be directly attributed to either a big crop or a small crop. What determines a big crop versus a small crop? It's not acreage! It's yield. The biggest determinant of yields is weather. If you have good weather, you get big yields; if you have poor weather, you get lousy yields. Here's a question I ask time and time again at seminars: "How many of you believe you can accurately predict the weather long term?" Each and every time, not one hand goes up. Typically, my next question is, "How many of you would be a better marketer if you never heard a long-term weather forecast in your entire life?" Nearly every hand in the room goes up. Face it, the weather guys have a very hard time predicting the weather more than five to seven days out. Sometimes they have a hard time predicting tomorrow's weather. If all farmers had all the money they lost and all the opportunities they missed because of some weather forecaster's incorrect prediction of an El Niño or a La Niña drought, there probably wouldn't be any need for a government farm program. Matter of fact, we might not even need ag bankers, because most farmers would be flush with cash. Okay, I might be exaggerating a little, but you get the point.

If you cannot predict the weather, you cannot predict crop size, which means you cannot predict prices. If you can't predict prices, then why base marketing on a price outlook? How can you make a marketing decision without an outlook? *Use strategy.* Plan for how you will manage the opportunity of a price rally. Equally important, plan for and be ready to execute your pre-planned actions, should prices fall. Be prepared and positioned to manage and benefit from whatever opportunity or risk the market throws your way. The next chapter will explain how this can be done.

Chapter 4
STRATEGY IS THE KEY

In the last chapter, we said that the only way to become a good commodity marketer is to learn to make decisions when you don't have information. This seems almost contrary to almost everything you have ever learned in life. You might even think it almost seems stupid. You can react to it however you like, but I have learned to accept it. In fact, I embrace it. The secret to overcoming the pitfalls of market information and market analysis is to learn to apply strategy rather than outlook to farm marketing. When I give speeches to farmer audiences, I tell producers they need to learn that good marketing is 90% strategy and 10% outlook. To be honest, I include the 10% outlook in the equation just to appease people and make them think I am not totally nuts. It is usually more effective when you ease people into a new concept rather than trying to convince them completely that their old ways were wrong.

How did I come to believe in the concept of strategy over outlook in the first place? When I graduated from Purdue University in agricultural economics, I was fortunate enough to get a job with one of the leading market advisory firms in the country, Top Farmers of America. I was hired as the assistant to a very well known and skilled market advisor. Remember, these were the days before PCs, so I was there to do the grunt work of hand calculating relative strength indexes and moving averages, gathering supply/demand data and taking phone calls.

Lo and behold, just a few months after starting this job (and remember, only a few months out of college), my mentor and senior analyst/boss decided it was time for him to venture out and start his own market advisory business. To say the least, this left a little bit of a void. We were publishing *FarmFutures* magazine, which provided substantial outlook information for all the major agricultural commodities on a monthly basis. That magazine was going to nearly 200,000 of the country's largest farmers. It was a pretty important audience! We also were publishing the Top

Farmer newsletter and the Market Insight charting service, which were sent to thousands of subscribers on a weekly basis. In addition, we had a daily hotline that we recorded every afternoon. That hotline was called by thousands of people. Remember, this was during the late 70s and early 80s when commodity prices had basically gone crazy, compared to the norm. Back then, people actually believed that we, as commodity market analysts, knew something and could be right. People flocked to seminars and bought our products aggressively. My point is that being the senior market analyst for this organization was kind of a big responsibility, especially for a 21-year-old.

Back in those days, all this commodity marketing stuff was pretty new, and there just were not many experienced advisors on every street corner with a sign out looking for a job. In fact, every major Wall Street brokerage firm was gobbling up these analysts and paying them big bucks. So this little ag firm I was working for found it rather difficult to replace my boss. As the months ticked by and I picked up his responsibilities, they began to realize that I was doing a pretty good job, and cost about a tenth of what my boss's replacement would cost. In the end, they never replaced my boss, and I ended up with his job.

So here I was, all of 21 years old, with three months experience and in a sink-or-swim situation. I had enough time during the day to figure out the supply/demand data for corn and soybeans. I had studied that while I was in college. I could keep up with all the writing and everything else, but when it came to the livestock and cotton markets, there just was not enough time or interest, and certainly not enough fundamental background for me to have a clue about what was going on. Sure, I had grown up working on a farrow–to–finish hog operation, so I knew how to give shots and castrate a hog, but I sure did not know what a high or a low slaughter number was or where to even begin to track meat demand. When it came to the cotton market, I could not even identify the crop if I saw it growing in the field unless the bolls were open and it was ready to be harvested.

You know what I found after a few years? My market advice was doing as well or better in the livestock and cotton markets than it was in the grain markets. All I was doing in hogs was looking at prices, realizing that we were at a moderately decent price level and putting a stop under the futures to get hedged just in case prices fell apart. I did not know if they were going to go up or go down, I just knew if they went down, I needed to be hedged. Lo and behold, hog prices would finally run out of steam in a bull market and start to fall, my stop point would be triggered and I would get short hedged. On at least two or three memorable occasions, I can recall that, when this occurred, I would broadcast on our daily hotline that the fundamentals still looked bullish, but we were hedged and we should stay hedged for right now. Prices would fall some more, and I would continue to report that the numbers looked bullish and that prices ultimately should turn and go higher, but for now we should hold our hedges. This would go on for weeks or months. In the end, on at least three occasions that I can recall over the years, I had people hedged for substantial (if not gigantic) gains the entire time that I was talking a bullish outlook and saying that prices could turn higher any day.

Putting it bluntly, my outlook was dead wrong. I was using the same information everybody else was using and predicting that prices would go in the same direction everybody else was predicting. But that didn't matter, because all the time I was short hedged and making money hand–over–fist for our clients. My outlook didn't matter; only my strategy mattered. If I had the right strategy in place, I could be dirt dumb on my outlook and still make money.

Think about it. How many times have you been wrong on the market? If every time you were wrong you made money, you probably would be rich. Having the right strategy in place allows you to make money regardless of whether your outlook is right or wrong.

The secret to becoming an advanced farm marketer is to move from basing your decisions on an outlook to spending your time and energy on developing a marketing strategy. Always know what you are going to do if the market goes higher, or if it goes lower. Always have a long-term plan in place. Do not allow yourself to make decisions or say things like, "Well I would sell, but China might buy," or "I would sell, but there might be a drought." Don't allow possibilities and "buts" to change your strategy. If prices are going higher, have a strategy to hold crop, delay sales or re-own crop. Also have a strategy in place to trigger sales if and when prices turn lower. Have a strategy to advance sales if that downturn extends itself lower. Always have a plan! This is "strategic marketing," not a marketing plan.

I do not like the term "marketing plan." It has been grossly misused. For years, too many extension people, professors and other market analysts have told people they have to have a marketing plan. To the best of my knowledge, not one of them has ever offered a plan; they just tell you to have one. Too often these plans center around picking a price level and then starting scale-up sales from that price level. What happens if the market never reaches that level? You are in deep trouble.

In a later chapter, I will go into more detail and give examples of how you can lay out a marketing strategy.

Please note: You just read the most important chapter in this whole book. This is a key concept that all good marketing revolves around – **STRATEGY**. Everything else in this book will help you either understand why strategy is so important, or apply the strategy.

Chapter 5
MANAGE OPPORTUNITIES AS WELL AS RISKS

In previous chapters, you have learned why I believe good farmers make poor marketers. We all naturally feel the need for information to make decisions. I offered a solution to this dilemma by suggesting you use strategy instead of outlook. This should be considered a rather strong suggestion; actually, almost an absolute.

Good marketing boils down to managing risks and opportunities. It is not predicting prices. It is not trying to outguess market action. It is not day trading. It is not pricing 1,000 bushels of your crop at the top of the market and then bragging to everybody about it while 99% of your crop is sitting in the bin as prices fall $1.00 a bushel. It is all about *consistency and maximizing the opportunities the market gives you.* That is managing opportunities as well as risks.

Let's back up for just a moment. Far too often, producers have been taught that they shouldn't be greedy and should sell at a reasonable price level (see Chapter 16 on breakeven). Unfortunately, farm prices spend way too much time at unreasonable price levels to make it possible to only accept a reasonable price level. When the market offers you $5.00 for corn, $12.00 for beans or $90.00 for cattle, you want to own something to sell at those prices. Preferably, you want to own a lot to sell there. That is your opportunity to get ahead or to catch up or to expand or to bring one of the kids back into the operation without taking on a lot of extra debt. It is your opportunity to make up for those years where prices never offered a good return. This is not greed; it is good business.

That is why I believe opportunity management is just as important as risk management in farm marketing. When the market offers you $5.00 for corn or $90.00 for cattle, you need to take advantage of that on a lot of your crop or herd, if not most of it; not a little of it.

At the same time, it is critical to know how to manage the situation when corn prices are at $1.00, hog prices are at $20.00

or when they are at risk of going to those levels. If you don't, you will dig a financial hole where only the richest of farmers can survive. If you throw in a bad yield year, an illness, or any other event, the combination could lead to a farm failure.

Farm marketing needs to be about consistency. You need to consistently see profitable returns. You need to consistently take out of the market as much as you possibly can within a reasonable risk parameter. And you need to consistently be a better marketer than your neighbors. Ultimately, in this farm economy, only the most prosperous will survive. You always need to be in a position to be able to afford a new piece of equipment more easily than the average farmer. You need to be able to write out that rent check more easily than the average farmer, and you need to be able to weather a bad yield year more easily than the average farmer. Agriculture is a marginal business and is always likely to remain a marginal business. Only those producers who consistently produce margins above and beyond the average will prosper and grow in the long run. Anyone at or below the average likely will not be able to remain on the farm.

I know this is not very warm and fuzzy, but I am not writing this book to make you feel good. I am writing this book to make you profitable.

Chapter 6
OUR BASIC STRATEGY

How do you manage both opportunity and risk? The easiest way to explain this concept is to give you an example. I developed the concept for what we call the Stewart-Peterson Marketing Program almost ten years ago. I could feel the need for this from the types of questions and reactions I was getting while doing marketing seminars. In the end, I simply joined together all the pieces of the puzzle to come up with an incredibly simplistic concept.

I started out at seminars by telling producers that if they did nothing more than learn to price one-third of their crop when they started to plant and one-third of their crop when they finished planting and possibly another one-third of the crop around the 4th of July, they would substantially improve their farm marketing. This example will work for corn and beans, but not for wheat producers. (Sorry, but there is no consistent pattern for wheat, so I can't give you a good example. Success in wheat marketing requires more sophisticated work.)

For a number of seminars, I coached people to just start doing more forward contracting earlier to improve their pricing. It seemed like a good concept, and a lot of people agreed with it. I did detect some concerns, though, from producers who didn't like the idea of having so much under contract before they were certain about the size of crop they would harvest. So I began offering an additional suggestion: "Whenever you are uncomfortable forward contracting, buy put options on the rest of it. (Puts give you the right to a short futures position. They lock in a floor price.) Then you are 100% covered against a downtrend in the market."

From the reactions and questions I was hearing, I knew I would need to go beyond this simple approach. Very often people are afraid to forward contract, because they are concerned they will not get a crop, or they are concerned they will forward contract at a low price or only a fair price when prices might get substantially better. This is especially true after prices have been to historically high levels. For example, in 1995-96 when corn went to $5.00 a bushel, many people had corn priced in the $2.60 to $2.80 range. When corn was at $5.00, they had nothing left to sell. Obviously, they had listened to too many bankers and extension people telling them not to be greedy and to just be happy with a reasonable profit margin. When they left $2.00 per bushel on the table, many of them swore they were never going to do that again.

As a result of this hesitation to overprice with forward contracts, I began advising seminar attendees to look at buying call options against the forward contract sales they had either made or anticipated making. This would give them the confidence to pull the trigger to lock in a price but still leave the upside open. After promoting this concept at a few meetings, it occurred to me that by buying puts, forward contracting and buying call options, a producer could be 100% priced if the market went down, and 100% open to take advantage of a price rally, should it occur. (Calls give you the right to a long futures position. They lock in a maximum price.)

To recap, let's say for example you forward contract 50% of your crop during the winter and spring before you plant, and you buy puts on 50% of the crop during that same time window. You are 100% covered on that crop if prices go downward. At the same time, let's say that you bought calls to cover 50% of your crop. Those calls offset the forward contracts you placed on 50% of the expected crop. If prices go sky-high, the gains on the calls will offset the majority of what was given up by having forward contracted at a lower price level. At the same time, the puts you bought on half the crop would more than likely become worthless, but you would have that crop available to price at the substantially higher price levels, should they occur. You are 100% priced in a down market and 100% positioned to take advantage of an up market. What more could you ask for?

This strategy certainly manages downside risk, and at the same time, it manages opportunities. What could be better?

Of course, as you likely have learned in life, there is probably no perfect spouse, child or marketing strategy. There is no free lunch. The strategy outlined above works well in any market that has a substantial decline. In a sideways market, the puts and the calls will lose money and actually give you a net lower cash selling price than if you had done nothing. How often are prices stagnant? Not often! That's why I believe it is worth giving up a little revenue in a sideways market to put yourself in a position to take advantage of a substantially higher market and to protect against a substantially lower market.

There are a few other drawbacks with this strategy. We found that often, farmers would buy the calls but then would not want to buy the puts or forward contract. This is a fatal flaw in about 70% or 80% of the market years.

We also found that buying the call options year-in and year-out was kind of an expensive undertaking when, in the majority of years, prices go down and the calls become worthless. Sure, if you bought calls every year over 10 or 20 years, you would be spending a lot on buying calls. But when you finally have a bull market, you would probably more than make back all the money you had ever spent. Buying calls every year can seem like you are spending money for protection a lot of years when it is not necessary.

The Stewart-Peterson Marketing Program is a good, basic marketing plan for starters. More sophisticated strategies can improve it. Possibly, only buying call options using trigger points at the start of a bull market can work just fine. More advanced option strategies on the call side can cut the cost substantially. There is a lot that can be done to make it work better, but the basic concept is very sound. It is strategic, it is not outlook-based and it manages both opportunities and risks.

Chapter 7
WHAT PRICES DO
IS ALL THAT MATTERS

All too often, well-meaning market advisors and farmers get too wrapped up in being "right" about the market. This means predicting where prices are going to go, knowing why they are going there and feeling all warm and fuzzy about how smart we are in our fundamental or technical analysis. This is all great if it works. Unfortunately, it does not work very often. All that really matters is what prices do.

I remember one time being totally bearish on the corn market. Based on the fundamental supply/demand analysis, I felt it would fall another 30 or 40 cents. It was fall, we had a big crop coming and there was no reason for prices to rally. All the fundamentals were bearish. I overlooked the market advisor's saying: "A known fundamental is a worthless fundamental." Lo and behold, some weather forecasters started predicting a drought for the following summer. It didn't matter that summer was still eight to ten months off. Corn prices rallied nearly 50 cents, and soybean prices rallied almost a dollar. To be sure, the rally did not last, and prices eventually receded to where I had expected them to go. In the meantime, I got my butt kicked. Was I right on my market prediction? Ultimately, yes. Did I lose money? Yes. Why? Because all that matters is what prices actually do – not what they are expected to do, or should do. Prices rose. It didn't matter why. It didn't matter that it made no sense at the time, and it didn't matter that prices ultimately declined again. What mattered was that prices rose enough in the short run to reclaim my equity in my hedges, knock me out of my hedges and put me in a weak position. I had been blind-sided, both in my equity position and in my mind.

My point is that being right on the market does not make you money. Taking advantage of price moves will win or lose the game. Remember that the only standard by which you should measure your success in the game of commodity marketing is dollars in your bank account. All you should care about is increasing the revenue per acre

What Prices Do Is All That Matters

your farm operation is producing. All you should care about is consistently producing good returns to your acreage and good returns to your operation, not whether you are right on your market predictions or analysis. You don't measure market success by your ego or your bragging rights at the coffee shop. You measure your market success by increasing your farm size without increasing your debt, or maintaining your farm size while decreasing your debt. Perhaps you could measure your success by having no debt and taking the revenues from your farm operation and investing outside of agriculture or in other forms of agriculture. We are talking financial success here, the kind you can smile about, your banker can smile about, your spouse definitely can smile about, the kind that ultimately lets you sleep better at night. Success means you can enjoy planting your next crop, because you are a successful farm marketer. You can confidently plant a crop with a realistic expectation of producing a profit.

Chapter 8
ALWAYS BE DILIGENT

Unfortunately, prices for all agricultural commodities spend far too much time trading near breakeven price levels. Too often, the best prices we see for the year are only marginally better than many producers' breakeven. That is why it's so important to always be diligent in your farm marketing. You need to be very defensive in your approach, because it only takes a small price move down to wipe out your profits and put you into the red. Prices always seem to fall easily and quickly. On the other hand, you can be almost blind and still catch a bull market. Once prices start going up, whether it be corn, beans or even wheat, if you missed the first 50 cents of the move, no big deal. There still can be more dollars on the upside to capture, even if you react slowly.

My point is that you have to react quickly to protect your profits on the downside. On the other hand, you can be fairly slow and cautious about responding to potential bull markets. After all, upside is, in theory, infinite. The downside is at zero.

Chapter 9
DANGEROUS YET USEFUL TOOLS

If you are like many producers, you may believe that futures and options are tools that are just too volatile and too dangerous for you to use. You may have tried them yourself and have gotten burned. Or, you may know of a farmer who managed to be ruined financially following his use of futures.

To start with, let me point out that I have never seen a farmer ruin himself financially by using futures or options for **hedging** purposes. I can tell you this after being in this industry for nearly 25 years and having literally thousands of brokerage accounts at our firm, almost all of which are farmer hedger accounts. I'm not saying it won't happen, or that it can't happen, but it certainly is not common. It is much more common for a producer to speculate heavily, with disastrous results. Often farmer speculators will make money on the markets and will start to feel invincible. So they start speculating on larger and larger amounts. As long as they're making money, it seems easy. They start believing that they have a real knack for this commodity thing. Then, in a matter of days, the market turns against their heavily leveraged position and they've wiped out all the money they've made, oftentimes in the hundreds of thousands of dollars. Too often, they end up deep in debt. In my experience, those who get buried by the market get buried by their own misplaced self-confidence and greed.

Getting back to the subject at hand, can futures and options be dangerous tools? The answer is yes. They are highly leveraged and highly unpredictable. But can't the same thing be said about farm machinery? I have spent many hours driving a combine, successfully harvesting thousands of acres of crops, knowing that the very machine I am sitting on could kill me. I've seen it first-hand. When I was working on a custom wheat harvest crew one summer during college, one of my friends/co-workers managed to get his hair caught in a shaft on the combine. If it wasn't for extraordinary efforts by the owner of our crew, he would have died that day. Yes, a combine can be very dangerous. But, far more often, it's a very useful and necessary tool.

How many people do you know of who have been hurt or killed by a PTO shaft? In our area, we have a lot of forage harvesting equipment driven by a PTO. The danger is ever-present, but it's darn hard to run this equipment without a PTO shaft. It is an essential item, but also a potentially deadly one.

Futures and options can be dangerous, but are absolutely essential and necessary to do the work of marketing. They absolutely can hurt you if you misuse them and do not respect them. Fortunately, in this case, they only hurt you financially. It is important to be knowledgeable (educated on their proper use), disciplined and always diligent when using any kind of potentially dangerous tool.

The fact that there are potential dangers in using these tools is not a good excuse to avoid using them. Either learn to use them properly, or hire professionals to do it for you.

Chapter 10
FARM SIZES ARE UP – BUT WHERE ARE THE PROFITS?

Throughout the country, farm sizes have been increasing. Part of it is just demographics. The average age of the U.S. farmer is 58. Over the last decade, thousands of farmers have "retired." Due to age, economics or health reasons, these older farmers have sold their farms or leased them out to other operators. Our average customer today is farming typically twice as many acres as they did 10 to 20 years ago. If you had asked those producers 10 years ago if they thought their farm size would double, they would have told you "No. There is no land available; it's too expensive; there is just not the opportunity." Despite that outlook, they have all grown and they have grown substantially. These customers are among the best farmers in the country. They are the best growers in the county, the best neighbors, and often the best friends. When land becomes available, someone comes knocking on *their* doors and offers it to *them*. Or they are the people who keep their ears to the ground. When they see an opportunity unfolding, they are ready and able to take advantage of it.

In the past, many of our customers had more machinery than they needed to operate their acres. It was natural for them to be able to pick up an extra couple hundred acres here or there, with the goal of spreading their machinery costs over more acres and increasing their net revenue. This was a good plan, and worked to a degree.

With today's increased market volatility, many of these same producers are farming twice as many acres but making no more money. The machinery costs, chemical costs and fertilizer costs have gone up substantially. Many of the production efficiencies and gains that have been realized by acreage expansions have been eaten up by increased input costs.

These larger farmers are now experiencing even wider and wilder swings in their income. They are not seeing consistently larger net incomes, just consistently bigger swings. One year they will make $150,000 and the next year they will lose $150,000. In just a matter of weeks, a big price slide in either crop or livestock markets can wipe out hundreds of thousands of dollars in profitability for even an average-sized producer. Being busy in the field, taking a vacation or getting sick can cost you twice as much when your operation is twice as big and you miss an opportunity.

The difference between the risks and rewards from good or bad marketing have never been bigger, and the potential for profit or loss gets larger every day as markets increase in volatility and farm sizes grow.

Your marketing skills or the skills of the professional marketers you hire have to rise to meet the ever-increasing challenges and opportunities.

Chapter 11
FARM MARKETING HAS CHANGED

Not too many years ago, I could tell producers at seminars that they could be good farm marketers just by doing a good job of forward contracting and planning their cash sales. When I graduated from college and started in this industry, I was told that farm marketing had changed, was continuing to change rapidly, and would be very different in the near future. After almost 20 years, however, hardly anything had changed, except that farmers tried a lot of different marketing tools and were generally unsuccessful using most of them. Over time, they were less willing to hedge, use options or even show up for a marketing seminar.

Then 1995-96 came along. We saw $5.50 plus corn and $8.00 plus soybean prices. We saw changes in the farm program. We saw South American production growing to levels where it became obvious they could out-produce the United States. Russia became a non-factor, while China became a big factor in the grain market picture.

Right then and there, the world changed. Farm marketing has never been the same since. Take corn as an example. Prior to 1996, you could pretty well count on corn to remain above loan level for all but a temporary window of time, when it typically fell just below the $2 level. Even in the most extreme bullish years, $4 was the top. So, we were working with about a $2 price range. At the time, that seemed really wide, and things heated up when we saw swings within that range. In 1996, however, we saw corn prices increase above $5. And to add to the volatility, with the new farm program, there is absolutely no price floor underneath the market. With really big crops, we could see corn go to nearly $0.

Do not rule out corn prices dipping near $0. It's not hard to even predict what events might occur to cause it. A few really big crops could do it. Just look at the hog or cattle industry to see what can happen. Those prices actually went below $0. At one point, cattle producers had to pay truckers to haul calves away and receive nothing for the animals – just a bill for trucking. Hog producers shot sows and had to pay the rendering truck to haul them off. Those are below $0 prices.

No one may believe now that prices could go to $0. But too often, it is what no one believes or no one predicts that really happens. Grain is more storable than livestock, but if you believe that something will not happen, you should be aware that it could happen.

Speaking of the unbelievable, I believe that, the next time grain prices go up substantially, they will likely climb to all-time highs. Corn at $5 will not stop the rally; it will go to $7. We could easily see beans in the teens again. Why? We have a larger demand base than we have ever had before. Demand is more in line with production than it has ever been before. While futures volume and open interest are up, this does not begin to compare to the amount of grain that is actually being handled. With a substantial market move, end users could scramble to cover needs at rates never seen before. In the past, knowledgeable people were nervous about buying corn at $3.60 or $3.80, because it had never gone beyond $4.00. Now they'll just be getting started when they buy at those levels. They may still be buying at $4.60 or $4.80.

My point is this: just because corn stopped near $5 the last time it went up, it does not mean it will stop there the next time. It used to be that corn traded from $2 to $4 – a $2 range. Now the actual range is $0 to $5. That's an increase in the range of 2½ times! The same is true for beans and the same is true for wheat. Look at the cattle and hog markets and the price levels we've seen in those markets. In the past 10 years, we've seen hogs trade as low as near $20 and as high as more than $85. Cattle have traded from the $50s to $100 or more.

This means that, yes, the world has changed. There is more opportunity than ever before, and there is more risk, too. But risk is the name of the game in the farming business. When you decided to become a farmer, you made a decision to face the risks thrown at you from Mother Nature and the markets. **The time has come to decide that you want to begin man-**

aging those risks. You need to use the available tools to shift all the economic risk you can shift.

Increased volatility, opportunity and risk are the reasons I believe so strongly that options are a necessary tool in the years ahead. It's nearly impossible for a producer to short-hedge any commodity by thinking that he can ride it for a few dollars one way or another, and have it ultimately be a good hedge. The upside potential in the markets is almost limitless. Maybe you and your banker can withstand the margin calls, but can you withstand the missed opportunity of a pure futures hedge that turns substantially against you? Instead, you could have purchased a put option (or a put strategy of some sort) that leaves the upside open and puts a floor under your prices.

Also, with these new markets and new levels of volatility, a futures trade brings substantial risk. You cannot go into the corn market, put a 5-cent stop on a position and have any reasonable expectation of retaining that position a week later, let alone three months later.

There were years and times where a person could almost do that – or at least they thought they could. Not anymore. Now, if you go into corn without giving it 20 or 30 cents room to trade, you are almost guaranteed to be bumped out before your hedge has a chance to work. The same is true in any commodity. I don't expect that to change anytime soon.

So yes, farm marketing has changed. That makes it necessary to move your marketing to a higher level, to change the tools you use and raise the levels of sophistication you employ. Otherwise, you will be leaving opportunities on the table and taking more risk than you can afford.

Agriculture has changed:

- Vertical integration of large grain companies
- Increase in farm sizes
- Globalization of markets
- Decreased Russian influence
- Increased Chinese influence
- Smaller margins throughout the industry
- Bigger price ranges
- Increased South American production
- Changes in government program
- Much more

Chapter 12
WHAT DOES IT TAKE TO MARKET WELL?

What does it take to be a good marketer? Obviously, strategy instead of outlook – I hope you agree with that by now. Assuming you have that concept down, what are the other successful ingredients? In no special order, I would say they are money, time, consistency, knowledge and discipline.

Money. If you don't have the money to be able to freely make decisions and implement those decisions, you are starting out behind the 8-ball. Too many options are not purchased because of a cash flow shortage. Too many decisions are delayed because farmers must make cash sales before they have enough money available to implement a marketing strategy. The old saying, "it takes money to make money," is true for farm marketing. You can do it on a tight budget, but you are somewhat handicapped.

As with any handicap, determination and skill can help you overcome many of the drawbacks of the handicap. But it takes a lot of extra effort in all the other areas to be equal.

For hedging with futures or options, as a general rule, plan to budget 8% of the value of a bushel for margins or premiums. Market volatility will heavily influence this. If you anticipate using futures or options on only 25% of the crop, then you might only need 2% to 3% of total crop value to cash flow your hedges.

Time. Being a good marketer requires time. You need the time to develop strategies and constantly review them to be sure you have all your bases covered for rallies and potential price declines. You need time to evaluate the various marketing strategies you could implement to find the least-cost method that has the greatest amount of leverage for the dollars invested. Most important, you need time to do this consistently throughout the year. Far too often, critical marketing decisions must be made during the planting and harvest seasons. If you are distracted by broken equipment or long hours in the field and are not making critical marketing decisions, the cost can be substantial. It is far too easy to develop a habit of saying you'll look at it tomorrow because you have to be in the field late tonight. Then tomorrow comes, the weather is good and you push hard again all day long and late into the night. You do this again and again and again, and before you know it, those positions you had on that were so profitable have given back all the gains, or the positions you planned to take never happened because the fieldwork took priority. Yes, you do have to plant and harvest a crop to have something to sell. But if you don't sell it well, can you afford to grow it?

I'm not saying the fieldwork should not take priority. It's pretty hard to market a crop you can't harvest. Remember that very few businesses of any kind can be successful without a good marketing effort! Allocate the time, delegate the responsibility within your farm operation, or hire an outside consultant to make sure the time is being allocated, not just when it's convenient, but all the time.

Consistency. Consistency is not really a resource allocated to marketing. It's more of a discipline. Consistency goes back to being willing to allocate the money necessary for marketing, whether the budget is tight or flush with cash. It's allocating the time to make a decision, whether you're busy in the field or not. It's checking your positions and making sure your strategy is comprehensive and covers all the possibilities.

Consistency alone can help you become a better manager. In my experience, I have seen many producers consistently implement what I would consider a poor marketing or trading approach and still be successful. Other producers have a very sound approach, but when they apply it inconsistently, they are almost never successful.

Knowledge. This does not mean knowing the latest fundamental news that you hear on the radio, the weather forecast, or what China might or might not buy. In today's sophisticated marketplace, knowledge means understanding deltas, thetas and a bunch of other terms and concepts that you may never have heard before or never expected to hear. You may never want to know about them, and certainly may not want to allocate part of your day to learn about them.

Knowledge is dramatically more important today than it was prior to the dramatic change in the farm market situation in 1996. In the old days, you only had to know a few simple forward contract alternatives. If you used futures or options, you only had to know how to buy or sell them. Knowing the difference between a price order and a stop order was helpful, but your broker likely kept you out of trouble on those.

The world is a different place today. As I've stated previously, I believe it's nearly impossible to do a good job of marketing in today's volatile markets without the use of options. The straight purchasing of options has become prohibitively expensive, due to the increased volatility. In our office, we have had to hire staff with substantial math and statistical backgrounds, as well as the aptitude to rise to the challenge of these new markets. We use sophisticated options analysis software which allows us to model how an option will perform over different price movement scenarios and changes in market volatility. We can compare a variety of different strategies by overlaying charts of the predicted option value models. It's not like the old days, where we just looked at the market, saw it was declining, and bought a put. Now we look at that as one alternative, then compare it to a bear put spread, a ratio spread, a fence strategy, a futures position with a sold or bought option against it, and a host of other alternatives. All this gets to be somewhat complicated and mind-boggling, but it's what must be done to find the most leveraged strategy with the least amount of risk and cost. It's constantly a challenge to weigh all of those conflicting goals and alternatives.

Certainly, some producers will eat this up. They will love the mathematical challenge and sophistication of using math to pick strategies, instead of using a gut feel and a guess that are all too often based on an outlook, and therefore doomed to failure. For many other producers, this is one layer of farming they may not have the aptitude, interest or discipline to want to learn or to do. In the end, those who have the knowledge and the ability to apply it consistently will be those who are the most profitable.

Discipline. After many years of implementing our farm marketing program for customers and having customers move up to our more sophisticated managed products in the Matrix family, we have found that no matter how much money, time and knowledge a person has, everything falls apart if there is no discipline. You need to have the discipline to develop your strategy, track the indicators you are using and implement your strategies when the signals are hit. It is far too easy to let an important market signal be hit and slip past you without acting on it. You may have a trigger point you decided on many months before and reaffirmed many times over, but when that signal is hit, you don't follow through because there is some market news that sounds just too tempting. You may hear that China may buy and you think you'll just wait a day. And you wait another day. When China doesn't buy, the market falls apart, and you regret your decision to wait. Now the price is too low, so you wait and hope it bounces back. But then it declines some more, and then some more, and then some more. In the end, despite a very well thought-out plan and strategy, you never pull the trigger to make the sale. The absolute best strategy is worthless without discipline.

(See glossary for explanations of the marketing tools.)

Chapter 13
THE STRATEGIC APPROACH TO MARKETING: AN OUTLINE

Being a strategic marketer requires a completely different mindset than what nearly everyone is used to. Almost every marketing decision is typically based on an opinion of where prices are going. If someone believes prices will go higher, they wait to sell. When prices instead fall lower, they're caught off-guard and don't make sales. If someone believes prices will fall, sales are made. All too often, if prices are good, the sales are too small to have a significant positive impact on the overall average price received. As outlined previously, all too often the confidence to sell a substantial quantity of crop doesn't come until there is such an overwhelming amount of negative news that the majority of the downmove has already occurred.

In strategic marketing, you always have a plan. You plan what to do if prices rally a little, what to do if prices rally moderately, and what to do if prices rally a lot. At the same time, you plan how to market should prices fall a little, how to market if prices fall moderately, and how to prepare for a dramatic substantial price decline. Your time horizon isn't over a day or even a week; it takes place over months and even over several years. You need to start planning the marketing of your crop years before it's planted, and consider how you've marketed that crop up to a year after it's been harvested.

Good strategic marketing is unemotional. It's calculated and planned. If done well, it can almost get to the point of being rather mundane and boring. If marketing is exciting, it means you haven't planned well, and there are surprises. Your only excitement should be at the end of your marketing window, looking at your average price, and seeing how much better it is compared to the average price received by the majority of farmers and in relation to what the market has offered. That is when you win the game.

The Strategic Approach To Marketing

Below is a basic outline of a strategic approach to decision-making.

Decision Criteria:
- Time of year
- Value
- Trend
- Cash flow
- How much to sell

If Sell:
- Cash
- Futures

If Cash:
- Basis (determines what tool)
- Carry (when nearby or deferred)

If Not Cash:
- Futures
- Options

If Futures:
- Stop point
- Option as stop
- Exit or roll/exchange for alternative

If Options:
- Puts/Calls
- Fence
- Spreads
- Ratio

Let's go through the outlined strategic approach step by step, and discuss it.

Your decision criteria is by far the most important part of the entire strategic approach to marketing. The tool you use to price is often not nearly as important as the timing of your sales or having the discipline to implement them. The first two items under strategy, time of year and value, go hand-in-hand. If it's well in advance of harvest and prices are historically low, there is no rush to sell. On the other hand, if it's well in advance of harvest and prices are historically high, you have to have the guts to pull the trigger and take advantage of those prices. You have to have a plan in place to buy back the board or use call options to confidently make those forward cash sales and not be fearful of missing substantially higher prices that might come along. This takes a great deal of discipline and willpower to sell well in advance of even planting a crop. It also takes a fair amount of research to recognize good value. Your gut feel is not likely to be a very good judge. We all tend to remember the extremes of prices. Good, solid research will tell you a realistic price from a good bet. To make it more complex, good value depends a lot on the time of the year. That research cannot be done over the life of the contract or the life of a crop year. It has to be done for each month preceding and post-harvest. This gives you the matrix of time over value, the first step to making good, solid analytical, unemotional marketing decisions.

The next step in a marketing decision is to determine the trend. If the trend is higher, then all efforts should be made to delay sales. If the trend is lower, every effort should be made to price as much crop as possible within your comfort zone and production capabilities.

Determining a trend is not easy. If it were, everyone in the world would be rich from trading commodities. The secret in determining the trend is to not look too closely. Years ago I was told that one of the best ways to determine the trend is to hold a price chart in front of an 8-year old and ask which way prices are going. That still may be one of the most accurate methods today. A long-term moving average, such as a 100-day moving average, is a start. Another is to watch how prices respond to news, as outlined in Chapter 18. Determining the trend is one of the toughest steps to nail down. Once you have done this, it's always possible that it will change. That's why managing all the possibilities of both up and down price moves is so important. You have to cover all the bases.

Cash flow is all-important. Far too many crop sales are made when money is needed, not when the marketing opportunity is the best. So plan months and years in advance when you'll need cash, and plan your marketing sales timing to be sure that cash is in the bank when the bills are due.

The next decision hurdle to overcome is how much to sell. For the first half or so of a crop that you're fairly confident you're going to produce, there are basically two methods to selling. If you sell very early at attractive price levels, large sales can be made as long as a solid plan is in place to defend those sales with a buyback strategy. A second alternative for the first half or so of the crop is to make a lot of rather small incremental sales in an effort to get a high average price. These sales need to be based on value relative to the time of the year. The more value you see, the more aggressive the sales should be. The closer you are to harvest or a seasonal downturn, the more aggressive you should be. The greater the opportunity for higher prices and/or a seasonal rally, the smaller the incremental sales should be.

It's all common sense stuff, but the more consistent and systematic you make your approach year-in and year-out, the less you'll be influenced by the latest news reports.

Once you've determined that you have value and there's a reason to make a sale, the next step is to decide how you're going to make a sale. Your two choices are either the cash market or the futures market. In the cash market, there's a whole list of forward contracting alternatives that are available through your elevators and co-ops. Under the futures category, you basically have the choice of either futures or options.

If you decide to sell in the cash market, then you need to determine if basis is currently attractive, and if there is a carry in the market. The current basis situation determines whether or not you want to lock basis or keep basis open. This directly influences whether or not you should choose a forward cash contract, hedge-to-arrive contract, a basis contract, or some other alternative. Watching the carry in the market influences whether or not you want to price nearby or deferred. This is very often overlooked, and can be a real bonanza for your elevator if you don't manage this properly. Oftentimes in corn, as much as 10 to 15 cents can be picked up by just properly managing the carry. The same is true in basis.

If you determine that you've already sold as much cash commodity as you're comfortable with, then the only alternative is to use futures or options.

If you choose futures, the next step is to decide your stop point, unless or if you're going to be a pure hedger. In that case, be committed to having an unlimited amount of funds available to maintain that pure hedge, and also be prepared to have an iron stomach to ride out whatever price moves come. I would venture to guess that 99% of the farmers do not have the finances and/or the nerve to pure-hedge.

If you're going to use stops with your futures, decide where you're going to place those stops before you enter the market. If you're going to place a relatively close stop, be prepared to be knocked in and out of the market often. You must be prepared to re-enter, or you won't be in the market for that big move. Typically for farmer/hedgers, a better approach is to give the market wide room to trade and place a wide stop. Typically the stop will have to be placed two or three times further away than you're comfortable. Statistically, using a wide stop is the only chance you'll have of getting in or staying in the market. A good alternative may be to consider using an option against your futures position as a stop. This makes the futures position a fixed-risk position, and makes it much easier to enter and hold.

In addition, if you choose to use futures, you should plan before you enter how you will exit the position or roll it, or exchange it for some alternative position. All these decisions should be made well in advance, before market activity and excitement influence your thinking.

If you choose options instead of futures, you must first determine if a put or call is appropriate for your needs. You should consider all the other strategies such as fences, spreads, ratio spreads, etc. Your goal in picking your option position should be to get the strongest delta at the least amount of cost. (Delta is the rate at which the option position moves relative to where the futures price moves.) A high delta number, such as 1, means that for every penny the futures moves, the option position moves a penny. A delta of .2 says that the option position will only move 20% of whatever the futures moves.

If you're using advanced option strategies where parts of them potentially have risks, you have to manage those positions, just as you would manage a futures position. Place stops or have the financing available to weather the storm, should adverse price moves develop.

Just like with futures, your option positions should be well planned out in advance, including how you plan to exit, under what signals, and at what point you will roll the options and/or exchange them for alternative positions.

Once you've gone through the process of making a strategy decision, you need to constantly (preferably daily and at a minimum weekly) review your positions. You should constantly reevaluate the amount of crop priced should the market move lower and, if you have enough crop ownership, should prices move higher. You should constantly evaluate how your current position (average selling price for your entire crop) will be affected, should prices make any significant move higher or lower. If you're uncomfortable with the net result of that average price, you need to invest more in the position to get a stronger delta, or you need to commit a larger percentage of your crop to forward pricing.

On the other hand, should prices rally and you see very little gain from that rally, you need to put into place a plan to either lighten up your hedges or reown crop that is sold, so that ultimately you capture enough of a price rally to feel good about your overall marketing.

This is a balancing act. In a sense, it's an art, but ultimately the more you work toward making it a science and a mathematical game, the more professional your marketing will be and the closer you'll be to marketing nirvana.

To do this process well takes an incredible amount of work and discipline. We often spend hours and hours evaluating dozens of different option strategy alternatives, trying to find the least cost, most bang-for-your-buck strategy. Once we find those strategies, we have to constantly monitor them to be sure we do not hold options too far into expiration when the time value erodes out of the option very quickly. Some option strategies can be placed with very little risk to their value for a short period of time. Once they near expiration, strategies can become very risky and costly. Strict adherence to rollover dates and/or critical price levels need to be monitored to ensure that strategies that initially looked attractive do not become very unattractive.

To add another layer of complexity, you also need to look at "loan deficiency payments" (LDPs) and the relationship of LDPs and crop insurance to the overall risk management structure you have in place. Oftentimes, the risk of an LDP disappearing becomes much greater than the cost of replacing it with a put option. At the same time, if you have an extensive portfolio of puts on, and still haven't taken your LDP, you are massively at risk of losing your LDP and your put values, should prices rally. Yes, you'd have cash to sell at a higher price, but you can leave a lot of potential revenue on the table. My feeling on this is that farmers have to quit trying to squeeze the market for every penny they can, and be willing to accept that they can't get the last drop of blood out of it. The old saying, "A bird in the hand is worth two in the bush," comes to mind.

Remember, you cannot have it all. You will never be able to sell all your crop at the top, and you will never be able to take your LDPs, exit all your short futures or take profits on all your puts at the absolute bottom of the market. No one is that smart or that lucky.

What you have to do is go for a good, strong average price consistently, year-in and year-out. You want to capture absolutely everything you can when prices are good, and avoid with a passion accepting low and unprofitable prices. It takes an awful lot of diligence, a lot of research, and even more discipline to be able to manage all the opportunities and risks that today's markets present you. That is the essence of marketing in today's farm economy. The tool is strategy!

Section 2
MARKET CONCEPTS YOU CAN USE

Chapter 14
ONE FACTOR DRIVES THE MARKET

In this day and age of satellite and cable TV, internet and satellite commodity information, it is very easy to be too well informed on the markets. This may sound a little bit odd, but it is very true. Far too often, many people believe a little more knowledge or a little more time will make them better marketers. I believe this is very untrue, and quite frankly, a waste of time. Let me get to the point. There is usually one major fundamental factor driving almost every market. It might be a large crop, or it could be unusually big demand. It could be a drought or a disease. Whatever it is, typically there is one major, dominant factor pushing prices either up or down. It doesn't matter if you are talking about corn, hogs or gold. This holds true. If you can identify that one fundamental factor and focus on that factor, you will be way ahead of everybody else.

The majority of people get lost in the proverbial woods, not seeing the forest for the trees. They listen to the radio, read their computer screens and get all excited about the latest export number, slaughter number, news of China buying or canceling a sale or whatever other normal insignificant news item is played up that day. Yes, I slipped that word "insignificant" in there purposely, because almost all of this talk is insignificant. You have to remember, you expect to read something every day. When you listen to the radio, you expect to hear news on the farm report. If you subscribe to an advisory newsletter, you expect them to include news. All these people make their livings by feeding you information. Yes, it is done with good intentions and with the idea it is of value to you. Quite frankly, most of it is just a distraction. It is just noise, getting in the way of you seeing the big picture. It is no different than the stalks and husks you have to run through the combine and spit out the back end of the combine in an effort to harvest that very valuable grain. When it comes to market analysis, the valuable grain is the one major fundamental driving the trend. All the other information out there is just husks and stalks meant to be trampled down, tilled under and ignored.

Just in case you do not completely agree with my point, let me give you an example to drive the whole concept home. For as long as you have been farming, can you remember a time from January or February all the way to May that you did not hear constant chatter about acreage estimates for corn and soybeans? Will acreage be up? Will it be down? Will there be a shift from corn to beans? For the amount of time and pages of print they devote to the talk of acreage, you would think acreage is one of the most critical factors in determining the price of a crop. Look back over the corn and soybean charts for the last 30 years. Compare where acreages climbed and where acreages dropped and see if there is a correlation between high prices and low prices. You will find that there is no correlation. Unless the U.S. government stepped in with some acreage set-aside program that artificially took massive amounts of acreage out of production, acreage is just not a factor. High prices are associated with bad weather, low yields and in rare instances, demand. Low prices, on the other hand, are consistently tied to big crops. Look at the numbers. Everybody gets all excited about a million-acre swing in corn plantings. What does a million acres translate into? At 150-bushel corn, it's 150 million bushels. First, if you look at the statistical variance the USDA places on its own estimates, a 150-million swing is well within the error tolerance they leave for their own predictions. Second, 150 million bushels can easily be added onto the crop by just one good rainfall moving across the central Corn Belt.

The bottom line is that, when we are talking about crop size, it will take about 500 million to one billion bushels difference in the corn crop to have a significant impact on the outlook for prices. Acreage does not do that; weather does.

What you should take away from this chapter is to remember that most crop news is worthless news; it is just chaff. Pay attention only to the one factor that is driving the market, and you will be a better marketer.

This will become more evident in the next chapter.

Chapter 15
BIG SUPPLIES, SMALL SUPPLIES – WHAT DO THEY MEAN?

If I told you the supply of a crop was really big and then asked you the direction of the price trend, what would your answer be? Most likely, you would say prices should be trending down, since supplies are large. That is a reasonable and logical answer, but unfortunately, it is very often the wrong answer. What I am about to say here is extremely simplistic, but if you understand and remember this concept and consistently apply it, you can stay out of a lot of trouble. In looking back over my experiences at predicting and trading markets for 25 years, some of the stupidest things I have done (such as being bullish on wheat all through a two-year downtrend), resulted from the fact that I ignored this simple concept:

– **If supplies are large but getting smaller, prices will trend higher. If**

supplies are large and getting larger, prices will trend lower.

– If supplies are small but growing, prices will trend down. If supplies are small and getting smaller, prices will trend up.

Put very simply, if supplies are increasing, prices are going to go down. If supplies are decreasing, prices are going to go up.

When you want to understand how the supply will influence prices, you should be looking at the ending stocks estimate, not the total supply.

The total supply level doesn't matter. What does matter is whether ending stocks are increasing or decreasing. Think about it a little. Let's say supplies are very tight but suddenly there are signs that demand has diminished or production was better than expected. That means the supply is not going to be quite as tight as expected. Buyers are less motivated to rush out and forward purchase, because they are less worried about a squeeze. Sellers are more

Big Supplies, Small Supplies

eager to sell, because their optimism has disappeared. Even when supplies are gigantic but are diminishing slightly, it leads to optimism. Buyers start feeling there is a reason to be a little more aggressive and purchase ahead, while sellers become a little more confident, a little braver, and feel they can hold a little tighter to their hope for a little better price.

Let's look at some past examples of this. In the mid 1980s, corn supplies were gigantic. Carryover supplies were well over 50% of annual usage. We could practically go without producing a crop for a whole year and still have enough to get by. If you look at the data, the minute we started reducing that massive carryover, prices started trending up from horribly low price levels.

Look at the extreme opposite. Take a year like 1980, 1983 or another crop year where prices went sky-high. As everyone perceives that we may completely run out of a commodity, prices climb higher, oftentimes at daily limit moves. The minute there is a crack in that optimism, where there are signs that the crop will be a little bigger or that demand is choking off, prices turn and collapse at a faster rate than they went up.

I call this the *light switch effect*. Markets go from being completely dark to completely light or completely light to completely dark instantaneously. There is no dimmer switch; it is all or nothing. If you look at how the psychology swings from one extreme to the other, you can see why it happens and how it happens time and time again.

I mentioned earlier the occasion when I was in the wheat market for a long time. I kept looking at how low supplies were and thinking that the downtrend would soon turn. I kept waiting to sell, figuring that the low had to occur any time. It only took a couple of years for the light to go on. I finally realized that, yes, supplies were small, but they had been consistently increasing because world production was adding to the world carryover situation little by little. Fortunately, I made that mistake many years ago and I have remembered it. It is painfully burned into my brain.

Big Supplies, Small Supplies

    I hope you can learn from my mistake. Don't focus on whether supplies are big or small to determine price direction. Rather, focus only on whether supplies are <u>perceived</u> to be increasing or decreasing. Knowing this will make you a substantially better market analyst and could keep you out of some major trouble.

Chapter 16
HOW IMPORTANT IS KNOWING YOUR BREAKEVEN?

A widely taught and commonly held belief is that you cannot do a good job of marketing your crop unless you know your breakeven price, and that all your marketing decisions should be centered on this.

In my opinion, managing your marketing based on the breakeven price can have a very detrimental impact on your overall marketing decisions and performance. I realize that a vast number of well-meaning extension seminars and a fair number of my competitors act as if breakeven is the anchor by which all marketing decisions should be based. Quite frankly, the market could not care less about your breakeven.

You need to manage your marketing to maximize opportunities and minimize risk. What if your breakeven price on corn is $2.30 per bushel and the best price the whole year is $2.20? Isn't it smarter to sell most of your crop at $2.20, rather wait for a price above your breakeven and ultimately be forced to sell it 50 cents or more below your cost of production?

Equally important, consider the missed opportunities in bullish market years if you sell at a "reasonable" margin above breakeven. Look back to the 1995-96 crop year, where corn prices went well over $5.00. After that market rally, I asked people during my seminars if they had a significant amount of crop left to sell close to $5.00. In all the seminars I conducted, only a few hands went up. In fact, most sold their remaining crop when the market hit what was viewed as an attractive/profitable price of $2.60 to $2.80. Why? They had been told by well-meaning bankers, extension people and others that you should be happy with a "reasonable" return on your farm operation, and that, when the market offers that reasonable return, you should sell.

I disagree. Farming is a tough business. You need all the money you can get, all the time. You deserve $5.00 corn when you can get it, because there will be times when you will have to suffer through $1.50 corn. If you market at a small margin most years, long-term you will go broke. It only takes a few years when that comfortable margin isn't available. That, combined with a few poor crop years, can suddenly put you out of cash.

Another reason not to use breakeven as a marketing benchmark is that you don't even know what your breakeven is until after all the crop is in the bin. Until you know your yield, there is no way you can accurately calculate your breakeven per bushel, or per acre. Sure, selling at a reasonable margin can be a good idea, but you must have a buyback strategy in place and be disciplined enough to implement that strategy. You have to be sure that if prices go substantially higher, you have ownership of crop and are in a position to maximize that opportunity.

At the same time, in years of massive supplies, the market may never approach your breakeven, let alone the price you're happy to take. Holding out for that elusive goal could well position you to end up with bins full of grain at the absolute worst price levels, and cash poor when the bills come due.

This brings up another warning. Marketing firms are misguided when they want to spend your time and money conducting a financial plan/accounting in an effort to get your "true cost of production" before they can help you with your marketing. Sure, knowing if you're a high-cost producer or low-cost producer is helpful in planning risk management. It also helps you to understand if the strategies you employ should be aggressive or conservative. However, you don't need an audited financial statement from one of the big eight accounting firms for this information. If you have high interest costs, are heavily leveraged or tight on cash, you need to manage your marketing differently than someone who has been farming for 50 years, has everything paid for, can let three years of crop sit in the bin without feeling a cash squeeze, and has all the well-maintained machinery he needs in order to farm until he retires.

Knowing your breakeven can be a little helpful, but do not get sucked into believing this is where good marketing starts.

How Important Is Knowing Your Breakeven?

It's typically where most bad marketing strategies begin.

Chapter 17
PRICE LEVEL: UNIMPORTANT!

Many farmers worry too much about current price levels when making their marketing decisions. Far too often, people will say, "I can't sell here! We're down 30 cents from where we were!" So they hold out, sometimes with disastrous results.

A good example is the soybean market. It's happened where prices have traded at $4 to $5 for two or three years, then rallied to $7. When that happens, it becomes impossible for a producer to think about selling at $6, because it seems way too cheap after seeing $7. Never mind that it's still $1 to $2 better than the past several years. That producer's perspective has changed.

The lesson to be learned here is to care about price direction, not about price levels. If prices are trending higher, you want to do everything you can to own crop or livestock and delay pricing as long as you possibly can within reasonable parameters of cost and risk management. If prices are trending down, you will want to sell everything you can as fast as possible. It doesn't matter how high prices are or how low prices are; you should follow those basic parameters. An old saying worth remembering is: "The trend is your friend."

Chapter 18
HOW TO USE THE NEWS

Is news useful? Yes, but not at face value. What do I mean? Let's say it's announced that China just bought a bunch of wheat. Should you get excited? Not necessarily. What you should care about is how prices respond to that news. If prices rally on the bullish news, it's a sign it was unexpected news and there is further rally potential based on this new bullish information. If prices do not rally on the bullish news, it's a sign it's old news, it's already built into the market, and prices cannot rally on friendly information. That would be a sign of a weak market.

If you look back at years and years of data, you will find a consistent pattern. It doesn't matter if you're looking at the market for corn, wheat, beans, hogs, cattle or widgets. Some of the biggest market-turning events came when prices failed to rally on bullish market news. That is oftentimes a deathblow to a bull market. The same is true on a bear market. Looking back into history, you will find many times when prices were trending steadily lower day after day. Then, some more bad news came out, but prices couldn't go any lower. Instead, they turned around on that bad news and went up. That is a very good, trustworthy signal that the bottom is in.

I love it when we have a bull market and bad news comes out. If prices can't sell off on that bad news, it tells me the bull market still has momentum and you can count on higher prices. That is a confidence-boosting event.

Bottom line: Don't listen to the news for what the news says is happening. Learn to listen for what the news says should be happening, then see how prices respond to that news. That is how you can use the news in a beneficial way.

Chapter 19
TAKE OUT THE EMOTIONS AND GET IT DONE

In a nutshell, it's best to make your decisions when the market is closed. This helps you to focus on the longer-term perspective, and avoid the emotions and anxiety of market action in the next few minutes or hours.

It's important to make marketing decisions well in advance of expected events and developments. Once you have decided what you want to do, get it done. Too often, people spend months planning to use a certain trigger signal to make a sale or a repurchase. Then, when the signal is hit, they look at the market and decide to wait until the next day's open to see if it opens higher or lower. If it's up, they decide to do one thing; if it's down, they will do another. My recommendation: when your signal is triggered, place your order on the next day's open, at the market. Just get it done! Waiting for the open doesn't tell you anything. All it does is give you more useless, confusing information.

Think about it. Let's say you have a sell signal triggered on Tuesday's close. On Wednesday's open you're going to go short futures. If on Wednesday the market opens lower, what does that tell you? Does it tell you it's a poor place to get short, and that you're selling in a weak, down market and getting a lousy fill? You could do better. Maybe. Or does it tell you that the market is very bearish, and the faster you get sold, the better, and you should get in immediately? Maybe.

Instead, let's say the market opens higher. Your plan was to sell. Should you wait? Will it go higher? Don't know. Is the higher opening a gift that's going to disappear? Again, you don't know.

The point is, no matter what the open, if you've made a decision to do something, just get it done. If the market offers you a better price to do it, be thankful. If you get a lousy fill, be thankful. It's probably a sign you really need to be sure to get your trade executed or cash sale made. You'll never know what is right until it's too late. A higher or lower open is not useful information that can be applied in a profitable and consistent manner. Statistically, if you consistently place your order immediately on the open, irregardless of a higher or lower opening call or opening price, on average you'll get the average. Sometimes you'll wish you had waited, and sometimes you'll be glad you didn't.

If you consistently do it, in the end you should be fine. If you inconsistently apply your strategy and try to outguess market action, you won't sell when the market is the most bearish (and you most need to be sold) because you won't like the lower open. When the market opens higher then fails to follow through on the higher open and collapses, again, you won't have sold because you will be waiting for a higher price. The market will continue to collapse, and you'll keep waiting, hoping it bounces back. You'll never implement the strategy that you planned to implement.

Just get it done. Don't get fancy; don't try to be too smart. The KIS principle comes to mind here – Keep It Simple.

Chapter 20
BANKERS: YOU CAN'T BLAME THEM!

Cash flow is one of the first problems many producers run into when they consider stepping up to the next level in their marketing. It's hard to implement marketing strategies or hire a marketing advisor well in advance of harvest when you may not have planted the crop yet, and are still sitting on last year's crop. As they say, money doesn't grow on trees. There certainly are a few producers in a strong cash position who never have to darken the doors of a bank. My goal is to help our customers become the majority of that group. But it doesn't happen overnight. In the meantime, most mortal men have to get some financing to make it all work.

If you have not already talked to a banker about financing for a hedging program or for hiring a market advisor/consultant, be forewarned. Odds are, your banker is not going to be too keen on the idea. And, you know what? You can't blame them. The majority of people who trade futures and options lose money. They don't have a plan, they don't implement a consistent strategy, and too many end up speculating – or should we call it "hedgulating"? It's a rare occasion for an ag banker to see checks returning from the brokerage firm. Their experience gives them good reason to be a bit suspect of your great hedging plan to lock in $3.40 corn when prices haven't been at that level in the last five years.

Their experience may also cause them to try and talk you out of hiring a marketing consultant to do your hedging and cash marketing for you. Over the years, too many people in my business have traded silver in a corn hedging account, have traded 100 contracts when only 20 should have been traded, or have had someone in and out of a position every three days, when it should have been left on for several months. Seeing that happen gives bankers reason to be a little bit leery.

Realize going in that, if you're going to do this and do it well, you're going to have to be a little bit of a pioneer. You will have to prove to your banker, your spouse, or whoever else may be judging you, that you can put together a good marketing strategy and have the discipline to implement that strategy. You can do the necessary research and due diligence to hire a professional farm marketing advisor who won't overtrade your account, won't "hedge" silver in your corn account, and whose true objective is to improve your revenue, not just line their own pockets.

It can be done and be done well. Your banker can play an active part. Just don't expect any of it to be easy at first.

Chapter 21
CAN YOU TRUST YOUR ELEVATOR?

I'm a little hesitant to write this chapter. It's difficult to talk about elevators overall, because there's such a big difference between one elevator and another, and between the ownership, the management and the help behind the counter of any individual elevator. Any generalization could be unfair.

Also, I have an inherent conflict of interest. In some ways, my business is in competition with the elevators for your farm marketing dollars. Ultimately, they always get the cash grain, but increasingly, they are making less and less money merchandising that cash grain and more and more money by providing marketing alternatives. Those alternatives sometimes may be in competition with the products and services my firm offers.

With all the disclaimers and niceties aside, can you trust your elevator? My answer is maybe, in some cases, a lot of the time, almost never, or it sort of depends. How's that for hedging my answer?

What keeps popping into my mind is that having your elevator do your farm marketing for you is kind of like hiring the fox to guard your henhouse. It can be a little risky. In the first place, an elevator's goal is to buy grain at the absolute cheapest price and the widest basis it possibly can. Your goal is to sell grain at the highest possible price with the best possible basis you can possibly get. As best as I can tell, those goals are complete opposites of one another. Inherently, this causes a bit of a problem. Putting that aside, let's say your elevator is just out to try to make a decent margin. Quite frankly, elevators haven't made a decent margin in merchandising grain for a lot of years. Look at the publicly traded ones. Their financial statements haven't been exactly great.

With merchandising margins being poor at best, more and more of the larger elevators (and eventually more and more of the smaller ones) are finding that the majority of their revenue comes from add-on fees for offering various marketing alternatives and tools. This could be a win/win, in that the elevator gets additional fees and you get additional tools. The elevators offer advanced marketing tools that are replacements for futures and options and/or may offer alternatives that the futures and options markets don't offer, all for a fee. It seems fair. They offer a service to you, and you should pay for it.

The good news is that, oftentimes, you don't have to pay for these services and fees until well after they're rendered, when you deliver your grain. It's just taken out of your grain check. You don't feel the margin call, you don't feel the option premium cost, you don't feel any of the pain until you get the grain check. It may be that the grain check is just smaller than it otherwise would have been, but it's still a nice check to receive. By using the tools properly, you could see your grain check increase. These marketing tools certainly have their place. They especially can be helpful to producers who are cash-strapped.

The problem arises when the elevator charges substantial fees above and beyond what is available in competitive marketplaces. In other words, they charge a 5-cent fee to give you a minimum price contract.

Meanwhile, your broker may have been able to accomplish the same option purchase for 1½ cents. If your broker tried to charge you a $250 commission, you would have screamed at him and said he was nuts. But at the elevator, since it's just a deduction from a grain check, you don't even notice, and it's convenient, so you easily pay the fee. If you do so knowledgeably and knowing your alternatives, that's OK. You're paying for convenience. But if you do so without understanding the true cost and the alternatives, you're throwing away good money and leaving revenue on the table that otherwise should be in your pocket.

Another problem I foresee long term is this: As the grain merchandising and elevator businesses become more concentrated, the more grain you'll be committing directly to that elevator early in the season under these types of marketing programs. That means the elevator can bid less for your grain later in the season. Ultimately, it becomes a control issue. When they gain control of the majority of the crop, they may in turn pay you a horrible price on the remainder of your unpriced crop.

No single firm has gained such a substantial market share that it can consistently employ this type of strategy – yet! But in coming years, it is more likely to occur. The more independent you can remain in your marketing, the better. The more grain you control, the more control you have over pricing decisions. The more producers who remain independent, the more competitive the marketplace will be to bid for the entire crop of all producers.

Another problem that we've witnessed on several occasions is: when dominant elevators have a substantial amount of contracted crop to be priced on a given date, somehow, mysteriously and suddenly, basis falls apart right around that window of time. Obviously, if they know they have a bazillion bushels committed to be priced and delivered based on a certain date, they have no economic reason to bid up the market on that date. But from a selfish standpoint, if that grain is going to be priced on that day and they can knock basis hard for that day, in essence, they just tripled the fees they charged upfront on those contracts to those customers. Suddenly, not only has the elevator made a 1 to 2 cent margin for merchandising your grain and 5 cents for the fee for the contract, they just picked up an extra 15 cents on the basis bump. They just turned a 2-cent margin into a 22-cent margin – a 10-fold increase in profitability. Not bad! Is the elevator your friend and can you trust it? In most cases yes, and most times the answer is yes. It's wise to know their motives and how they're making their money. Look beyond the obvious and don't be too quick to succumb to the convenience they offer.

It's no different than the combine, the PTO or the chainsaw. The tools they offer and the motives behind them are very useful and may even be very well intentioned. But you need to be a knowledgeable consumer, diligent in your use and application of the tools and offers provided. Always be a little leery that there might be a fox guarding that henhouse.

Chapter 22
DON'T TRY TO OUTGUESS THE REPORT

I've seen this for 25 years, month after month, time and time again. The USDA is about to release a report. Everyone calls to ask what we think the report will say. They sometimes don't like what I have to say – that I believe this is a waste of time and an inherently flawed approach to marketing.

In all the years that I've been in this business, there has only been one year where I believed I had a clear idea of the crop size, contrary to what the USDA predicted. In that year, I stuck with a number that was rather contrary to the USDA prediction month after month, until the USDA finally revised their numbers to match mine, many months after harvest. I was confident in that number because of the broad base of customers we have across the country, and the yield reports they provided. But that's once in 25 years. The rest of the time, give or take a little bit one way or the other, the reports are what they are. Unless you think you are gifted with some brilliant knowledge that the rest of the world doesn't have, what makes you think you can outguess the USDA number? USDA crop reports are funded by the federal government. They have more resources, more people and more experience than any organization in the world to conduct surveys and make their crop estimates. No one could do any better.

What about those reports that surprise the market? By definition, a surprise number is one that's unexpected. Often it's unexpected just because it makes no sense, based on the common knowledge that's out there.

It's not necessary that the number be accurate – just a surprise. Perhaps the number seems kind of stupid at the time, based on crop conditions. It doesn't matter though – it's a surprise, and the market moves based on it. A surprise number has half a chance of being bullish and half a chance of being bearish. Unless you feel lucky, betting on a report one way or the other is, in my opinion, kind of a foolish undertaking.

Don't Try To Outguess The Report

I believe you should go into a report as if it doesn't exist, and come out of the report as if it didn't occur. Usually within a couple days after the report, the market goes back and resumes the original trend that was in place, and all the hoopla and news surrounding it turns out to be nothing but a distraction. If it does change the trend of the market, your strategy should be in place to respond to the trend change and/or the change in supplies. That is a fundamental change. The fact that it came as a result of the report shouldn't be a big factor.

My advice is to not get excited about these types of events. Anticipating what they might say may lead you to outlook-based trading rather than strategy-based decision-making. Just ignore the whole thing, stick to your marketing strategy and enjoy your farming. Discipline is the key.

If you really can't resist the urge to try to outguess the USDA report, I suggest you bet your buddy on it, but do your best to keep the whole effort separate from your farm marketing.

Chapter 23
DON'T BE A SHEEP

The point of this chapter is to strongly encourage you not to be a follower. Don't do what everyone else is doing. You need to be a contrarian. In other words, when everyone else is bullish, you need to find the reason to be bearish. You don't want to carry this concept to extremes because, at some point, there is some value to having everyone agree with you. That means they are driving the market in your favor.

Remember the original concept: a known fundamental is a worthless fundamental. Remember Chapter 2 where we talked about why, when everyone is bullish, prices can fall because all the reasons for prices to go higher are already in the market. These are the same reasons you want to be a contrarian. When everyone can argue a good bull story, it's over and it's time to be a bear.

A typical occurrence (funny, yet sad) is that, far too often, producers call our office to ask what everyone is doing. We'll tell them that everyone is selling corn right now. They respond by saying they will sell corn. That is the absolute worst way to make a decision. If you find that everyone is selling corn, it's probably time for you to buy it. The majority of people want to sell after prices have gone down long and far, and they're just desperate and give up.

Let me fill you in on a little secret. Let's say we call 100 clients to recommend they sell wheat. If 70 or 80 of them follow the recommendation, you can almost be guaranteed it's wrong. Why? All the fundamental news that drove that many clients to agree to sell wheat is already built into the market and the move is over. On the other hand, if we called that same 100 people with the same recommendation, and only a couple followed it, you can almost bet that we will be right – short hedging where we recommended was the smartest thing a person could have done. It's ironic, but it's true. I've seen it happen time and time again, year after year. Marketing and market analysis is one place where "fitting in" and getting along with everyone else doesn't pay. You have to be an independent thinker. Certainly, don't follow the flock and don't be a sheep.

Chapter 24
OPEN YOUR TIME WINDOW

No, I'm not talking about airing out your house and no, I'm not talking about tinkering with Microsoft's PC program. I'm talking about your marketing opportunity window. I recently had the opportunity to read a book written by Donald Chafin and Paul Hoepner, entitled "Commodity Marketing From a Producer's Perspective." It's a very well-written book that covers all the basics of farm marketing. One of the major premises they promote in their book is to use a 730-day window for marketing. Their goal in promoting this concept is to encourage producers to start thinking about marketing a year in advance of harvest, and up to a year after harvest. This is a very good point, especially considering that they first wrote this book in 1989. I agree with the concept, although I think it should be carried even a step further.

If you do some long-term market research, go back and look at December corn charts for the last 20 or 30 years; it will almost make you cry. What you'll see is that when December corn futures start trading nearly two years in advance of harvest, it is often at very profitable price levels. Very often, except for extreme drought years or other wild bull market years, the highest price for the whole year is seen near the time when trading for that contract commenced. From there, it's just a long price slide into expiration, punctuated by an occasional rally to give some false hope to the would-be bulls.

In my view, you should look at a minimum 730-day window for your marketing. A 1,000-plus-day window is prudent. You actually should start looking to market your crop the day the new crop contracts start trading. That is oftentimes where some of the best opportunities lie.

Chapter 25
CARRYOVER STOCKS: MARKET SHOCK ABSORBERS

Grain stocks act as a shock absorber to market prices. When there is a large carryover of stocks, they act as well-dampened shock absorbers. No matter what kind of bullish or bearish news is thrown at the market, it typically has only a minor impact because it can't dramatically change the ending stocks situation. When stocks are large, it takes an outright crop disaster or unprecedented demand to cause a significant change. The large carryover supply acts as a shock absorber and dampens the effect of the news and events.

In a year when carryover is low, it's like driving a car with no shocks. Every hole you hit (news event, weather, etc.) has a jarring impact on the market. Prices react quickly and rebound quickly. There is no dampening. Markets are exciting, volatile and things happen fast.

With a low carryover, supply/demand numbers can change dramatically. With an extremely large crop and any moderation in demand, carryover supplies can double or triple. No matter how you slice it, that news is extremely bearish. At the same time, without the dampening effect of a big carryover, any crop problem will directly impact prices. With tight ending stocks, it can be nearly impossible to further reduce carryover, so prices must go higher to choke off demand. This can yield phenomenally bullish potential.

Chapter 26
BUY STRENGTH, SELL WEAKNESS

Many times when we've had a bull market in grains, I've heard people say that they didn't want to buy into the market because it was up. They felt prices were too high and they wanted to wait for a setback to buy in. That setback may never come, or comes so quickly and is so shallow that the buyer never enters the position.

In bear markets, I see the same thing happen. Everyone wants to sell while the market is climbing. Once the market tops, no one is willing to sell on the way down. Prices are always too low compared to where they were five minutes ago or five days ago. As a result, everyone always wants to wait for an upward retracement (rally) to sell. Just as in the bull market, the upward retracement never occurs, or if it does it is so shallow or comes so quickly that the short position is never entered. Quite often, even if a decent correction occurs, the person looking to enter a position loses confidence, worrying that the upward correction may actually be a trend change rather than just a correction. The lack of confidence leads to inaction.

By being selective about how you enter the market, you select the worst trades and miss the best trades. This is true for speculators and hedgers. If a market is going down and you're looking to buy, is that price decline a sign of strength? No. In fact, the more it goes down, the greater the risk it will keep going down. In this case, you don't want to buy at all. If you buy in an upmarket that is strong, you're buying when the market is exhibiting behavior you want to continue. Trade with the trend.

When making marketing decisions, don't let the little decisions get in the way and distract you from making good big decisions. Pay attention to the major trend and the single major fundamental that is determining that trend. Once you determine the trend, position yourself to take advantage of it. The sooner you do it, the better off you are. Don't let the possibility of a slightly better entry point get in the way of making the global decisions that make or break your marketing.

Chapter 27
MARKETING: AN INPUT EXPENSE

"Freedom To Farm" legislation, written into law in 1996, brought dramatic changes to U.S. agriculture. One of the most positive changes has been to allow you to plant the crops you want. Unfortunately, another change is what some have coined as "Freedom To Fail." Before the revisions in the farm program, low prices were at least partially offset by government deficiency payments. This meant that if farmers were fortunate to price at a high price and get good yields, they prospered. If they did a poor job of marketing and sold their crop at a low price, the government deficiency payment rescued them and allowed them to continue to farm. The new farm program does not offer this safety net. As a result, the low prices that will occur in the coming years will have a far more dramatic impact on farmers than ever before. Several years of low prices can have a devastating impact on a farmer's financial situation. The evidence for this is clear, according to data presented at a marketing conference in Michigan by David Kohl, an agricultural economist at Virginia Tech. Kohl's data shows that, on average, Midwest farmers would have lost money, if not for farm program payments. Only the above-average producers have been profitable.

IT'S TIME FOR A CHANGE! Farmers need to start looking at risk management/marketing the same way they look at managing the remainder of their crop operation. No farmer would consider planting a crop without budgeting and applying fertilizer or chemicals. However, few farmers have a budget for marketing. No one doubts their return on investment in fertilizer or chemicals. For all but a few, the process of writing a check for a margin call, an option premium, or even the process of signing a forward contract is a painful process.

A look at a few numbers may convince you to start to budget for marketing. The numbers will show the cost and reward of various management decisions. Obviously, we have to make some assumptions. Your operation will vary from the numbers used for illustration purposes.

Fertilizer: Assume you would lose approximately half of your yield if you applied no fertilizer. For corn at 150 bushels per acre, that would amount to a loss of 75 bushels of corn. Assuming a $2.60 selling price, it amounts to $195 per acre. If you assume $50 per acre input cost for fertilizer, there is approximately a 4:1 return for applying fertilizer. In other words, for every dollar that you invest in fertilizer, there is a $4 return.

Chemicals: Without pesticides, assume a loss of a third of your yield. On 150-bushel corn, you would lose 50 bushels of corn. At the same $2.60 per bushel price, that is a loss of $130 per acre. If you assume a $25 per acre chemical cost, the return on your investment in chemicals is roughly 5:1.

Marketing: There are a great many ways in which producers can go about their marketing to better manage risk. One of the most simplistic is to use puts to forward price part of the production, and forward contracts to price the remaining percentage of production. Crop covered with forward contracts can be covered with calls to take advantage of a market rally should a substantial price move occur. Assuming a cost of 15 cents per bushel for the puts or calls, 150 bushel corn would equate to a $22.50 per acre input cost. If you capture 50 cents of a price move (that's the minimum December futures move from 1993 to 2003), the return will be $75 per acre. That is more than a 3:1 return ratio. At the same time, if a drought were to develop and a substantial rally occur, calls could return as much as $2 per bushel, for the equivalent of $300 per acre. That is a return ratio of well over 10:1.

The simple analysis above shows that returns to $1 invested in marketing compare well to the returns of other input costs. However, producers typically find it very difficult to write the check for the marketing expenditure. A field without proper fertilizer or pesticide application is obvious to its owner and all the neighbors. Every day that you drive by that field, you are reminded that you should have done a better job.

On the other hand, marketing expenditures seem more optional. There is a chance prices may go sharply higher, and having done nothing will be the best alternative. In the past, if prices went lower, the farm program was there to rescue you. The cost of taking that risk is much greater now, and will be much greater in the future. Missed price rallies can be just as expensive as pricing and selling too low. The extra income obtained from pricing properly in a bull market may be critical to the long-term survival of many farms.

It's time to change your way of thinking. Budget for marketing, invest in marketing, and expect a return on your marketing. Lay marketing plans out well in advance. Use well-planned strategies rather than trying to outguess price moves. Without a marketing budget, cash flow and emotions may limit you from following through on the best laid plans.

University of Illinois research published in 1992 shows a dramatic difference in profitability between farmer producers who believed hard work would be rewarded and those who believed they needed to concentrate on financial planning. Those who focused on financial planning were much more profitable than the "hard workers."

With changes in the farm program, it is very likely that farm sizes will grow in the future, resulting in more hard work for more producers. If that hard work isn't accompanied by a greater emphasis on management (including marketing), then increased volume, combined with the increased volatility of the markets, will have devastating results.

Chapter 28
OPTIONS ALWAYS LOSE MONEY, SO WHY USE OPTIONS?

It is not uncommon for farm marketers to go through a cycle of basically making cash sales when cash is needed. Then they learn they are too often making sales when everybody else is making sales, and prices are depressed. Next, they venture into the realm of forward contracting. This certainly has some advantages, but it is hard to forward contract the majority of the crop before it is produced. There are always production concerns. Moreover, there is the fear prices might go higher (a bigger concern for many producers); many will wish they hadn't done any pricing at all.

After a producer gains experience and confidence with forward contracting, he or she will often move to the next level, and will realize the flexibility that the futures market offers. They start using futures. Just as it is difficult in life to find the perfect spouse or raise perfect children, it is also difficult to find the perfect marketing tool. While futures offer flexibility, they also are highly leveraged, and can be difficult to get into and stay in for long-term trends. Market volatility has a way of bumping people out before they can make money.

Many futures users eventually move to the next level and decide to learn about options – puts and calls. Unfortunately, options are no different than any other marketing tool producers have tried – they are imperfect. When I give speeches, I often ask the attendees, "How many of you have used options, lost money, and swore you would never use them again?" I can safely say the majority of people in the room raise their hands with big smiles while shaking heads. The fact is, though, that a major source of frustration producers express in using options and the reason for their general lack of success is directly tied to the type of options they buy.

See if this scenario looks familiar: You call up your broker looking to buy a December corn put. It is May, prices are looking pretty good and you are fearful that, if the acreage numbers are as large as they are talking about, ultimately prices will be much lower by fall. You ask your broker for an at-the-money December put price. He tells you it is 19 cents per bushel, which amounts to $950 on every futures contract of 5,000 bushels. You were looking to do 100,000 bushels, or 20 contracts. Multiply that times $950 per contract, and you end up with $19,000 in total option cost, excluding the transaction fees. Your reaction is "ouch!" You ask how much an out-of-the-money option costs. The next strike out is 16 cents – still too expensive. Then, if you are a typical producer, you may ask, "What about a September option instead of December?" They are only 3 cents cheaper. Your yields weren't so great last year, and fertilizer costs were sky-high. You are feeling a little pinched, so writing a check out for $10,000 or $20,000 for put options seems like an unmanageable expense. You decide that maybe a July option would at least carry you through until summer, so you ask your broker what an at-the-money July option costs. It is still too expensive, so you settle on a one strike out-of-the-money July option for 9 cents. You feel happy, because you lowered your cost to under a dime.

Here's the problem: you now have an option that will expire in June, way before the summer market volatility occurs and way before you know whether you have a crop or not. Yes, it will protect you through any kind of spring price slide, but only for a few months. You also purchased an option that is out-of-the-money, so prices have to fall 20 cents per bushel on corn before you even begin to make anything at expiration. You are fighting an uphill battle before you even start the war. Next, since you bought a July option, you are already into the time window where the option premium will start to diminish quickly because of time value erosion. Options lose their time value the quickest in the last 60 days before expiration. With an option in this time window, you can be right on the market direction and still end up losing money. That is really discouraging.

The problem is pretty clear. What is the solution? One of the simplest solutions is to not buy the cheapest option you can get by with. Think of it like farm machinery. When you go out to buy farm equipment, you do not buy beaten-up, run-down junk that was of poor quality when it was originally manufactured. Rather, you will typically buy the green or red equipment that has proven to be reliable. It is certainly painful to pay for it, but you know it is going to perform when you need it.

Buying an option should be no different. More than likely, it is going to be painful to pay for it, but if you invest in it properly and manage and care for it properly, it will perform for you.

There is an alternative to paying big bucks for options. Just like buying machinery, if you are a smart purchaser, you can save some substantial money. As with machinery, it takes extra time and effort to be a smart option shopper. In the case of options, it takes effort, education and sophistication to save money. It also takes more discipline. The only real practical way to save money on buying options is to use more advanced option strategies such as bear put spreads, ratio spreads, fences and a host of other types of sophisticated option strategies. (See glossary for explanation of these tools.) These tools can be very useful, but each one has certain limitations. Some have increased risk. Almost all require more management and discipline than almost any other marketing tool out there. They are a little like riding a spirited horse or driving a powerful car. They have a lot of potential, they can be very satisfying when used properly, but if mismanaged, they can hurt you.

Returning to the original premise of this chapter, yes, options can lose money a lot of the time. However, in today's wild, volatile markets, in many cases they are almost the only marketing tool that works. Unless you have an iron-lined stomach and a bank account that would make Bill Gates blush, futures can be too unruly in today's markets. Cash markets oftentimes do not offer enough flexibility. So, options become almost a necessity.

Chapter 29
HOW ABOUT USING TECHNICAL ANALYSIS?

With the advent of all this fancy computer technology that has come along in the last decade or two, a host of information is right at your fingertips – moving averages, relative strength data and stochastics, just to name a few. Great? Did you notice the question mark instead of an exclamation point? There is a reason. The fact is, the majority of these tools are of questionable value.

Much of the big commodity fund money is traded off of long-term moving average based systems, but they are much more sophisticated than that. I don't know of anyone that has used stochastics to actually trade and make money. I personally have done research on the relative strength index (RSI), and have watched it for 25 years. What you find in practice with the relative strength index is absolutely the opposite of what the general public and published data would tell you. When the relative strength index is high, normally the belief is that the market is "overbought" and due to sell off. So, a person would short the market. That may be true if you short the market for a day, maybe two or three days at most. But the data I've looked at over the long term shows that a high relative strength index shows a strong uptrend. If you want to trade with the trend, you should be buying the high relative strength index and not selling it. It's true that high RSIs are almost always evident at market tops, but the high index is also typically present for the last third or more of the entire upmove. If you sell a market due to high RSI, you're selling into a strong uptrend and are likely to be buried in losses. Considering the level of knowledge with which the majority of people try to use it, the relative strength index is likely to lose money consistently.

My point is that you should definitely understand the limitations of some of these tools. It's kind of like having an $^{11}/_{16}$" box-end wrench in your toolbox. It's handy to have there, but you certainly can't plan the maintenance of your entire fleet of farm

How About Using Technical Analysis?

equipment around that one wrench. It's pretty limited in its scope and its ability. Most of these readily available tools are exactly the same. They have some usefulness if applied properly and consistently, but in and of themselves, you aren't going to rebuild an engine with just an $^{11}/_{16}$" wrench. As a matter of fact, you might do an entire rebuild and never even pick the thing up unless you need it as a pry bar.

Chapter 30
INFLATION AND OTHER INSIGNIFICANT NEWS

If you farmed through the 1970s and early 80s, you heard lots of farm news about inflation and how it would drive ag prices up. I find it amusing how people get so fired up talking about things that are so misguided. Look back at those years. When we had a bull market in 1973, corn went to near $4.00. When we had another bull market in 1974, corn went to $4.00. Again, in 1980, when we had a bull market, corn went to near $4.00. During that same time window, the price of a car at least doubled, and the price of just about everything, including many farm inputs, doubled or tripled. However, corn prices didn't double, bean prices didn't double and wheat prices didn't double. In fact, it is now 30 years later, and they still haven't doubled. Cars cost four or five times today what they did back then. Certainly tractors and combines cost way more, but agricultural commodity prices are still basically the same. What does this tell you?

In high school economics, we learned that the definition of inflation is too many dollars chasing too few goods. Fortunately and unfortunately, U.S. commodity producers are far more capable of increasing yields and increasing production beyond the market's ability to handle it all. The good news in recent years is that we are seeing more of a balance, and still, production capability is always there to offset demand. Therefore, there are never too many dollars chasing too few goods (bushels of crops). That only occurs in short time windows when there is a perceived commodity supply shortage. That is not inflation; it is simply the law of supply and demand.

While inflation has not been a big factor in the news for the last decade or so, it very likely will become more of a factor as we move through the 21st century. Everything goes in cycles.

But, as in the past, it's highly likely that inflation will affect your input costs more dramatically than it affects the revenues you receive.

Chapter 31
WHAT ABOUT THE STRENGTH OF THE DOLLAR?

In the previous chapter, we talked about the misplaced importance of inflation in the past. As focus on inflation diminished, the focus shifted to the strength or weakness in the dollar. It has become a bigger factor in a market analyst's toolbox and, therefore, has become a big factor in the daily news media. While I will admit that the strength or the weakness of the dollar affects commodity prices, I should point out that, in my opinion, the focus on this is way overdone. There may come a day when the strength or weakness in the dollar is the primary driving major fundamental factor behind a market move. If so, get excited about it, pay attention to it and ignore all other information. Until that happens, I suggest you pay little or no attention to it. Here is why:

Factors like the strength or weakness of the dollar affect the extent of a market move, not the direction. In very simple terms, if you are looking at a wheat price that is rallying to near $4.00 and the dollar is weak, maybe wheat will rally to $4.10 or $4.20 instead of $4.00. Or, if the dollar is strong, maybe wheat will only make it to $3.90 instead of $4.00. The point is that wheat is rallying. The question is how far the move will go. I don't know how smart or how good you are at predicting wheat prices, but I sure do not think I can predict wheat prices within 10 cents one way or another.

The only thing a weak dollar does for a bull market is help the bull market go a little further than it would otherwise go. It makes it a little easier for it to go up. The only thing a weak dollar does in a falling market is make that market fall a little slower, or maybe not quite as far.

My belief is that you need to focus on the direction (the trend) of prices and not the extent of a move. It is almost impossible to know how far prices will go. You'll become a better marketer when you get out of the habit of trying to predict the extent of a market move. Just believe in the trend and

What About The Strength Of The Dollar?

follow it for as long as it lasts. Do not try to outguess where it will stop.

While this chapter is focused on the strength or weakness of the dollar, <u>the same concept applies to a whole host of other minor fundamental factors that contribute to the extent of market moves rather than direction</u>. Remember to focus on that one major fundamental driving the market. Let the price action and momentum of the market tell you what it cares about and what is driving it, not what the latest farm news report says.

Chapter 32
BELL CURVE YOUR SALES

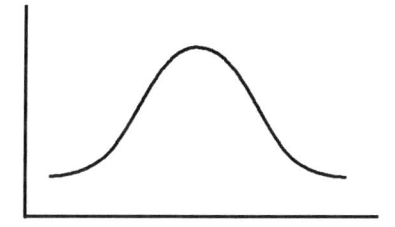

Remember back in school when test scores were graded on a bell curve? A graph of a bell curve looks just like a church bell sitting on a platform. The graph starts out low, gradually slopes upward, the curve steepens, peaks, falls off fairly steeply, and then gradually slopes back to the axis.

When test scores for a school exam are graphed, a bell curve of results usually shows a few very low scores. Then the curve slopes up dramatically, because so many students fall in the middle with average scores. The curve peaks at the average, then drops back off sharply, showing a few students with near-perfect or perfect scores.

The same concept can be applied to your crop marketing. Let's say, for instance, that today you are looking at December corn futures trading near $2.30. You feel this is an unacceptable price, but you are also fearful that the upside potential is fairly limited. You realize that prices often have traded as high as $2.60 or $2.70. With just a moderate drought scare in July or August, that kind of objective is obtainable, but also a bit of a reach. You know that, as the market rallies, you should sell some crop, but it is difficult to know how much and how fast. If you bring the bell curve in and apply it to your marketing, you may find some comfort and discipline.

Here's a rough idea of how to make practical use of a bell curve for your marketing. The table below gives an example of how to scale-up sell within an expected price range. As you can see from the graph and the data, you start to sell lightly as the market begins to rally. As the rally accelerates, the number of bushels you market also accelerates to larger quantities. As the market begins to pass the mid-point of your expected price rally, the number of bushels you sell begins to diminish.

Bell Curve Your Sales

The following example assumes a total of 30,000 bushels to be marketed.

Corn Price	Bushels Sold
$2.30	1,000
$2.35	2,000
$2.40	3,000
$2.45	5,000
$2.50	8,000
$2.55	5,000
$2.60	3,000
$2.65	2,000
$2.70	1,000
	30,000

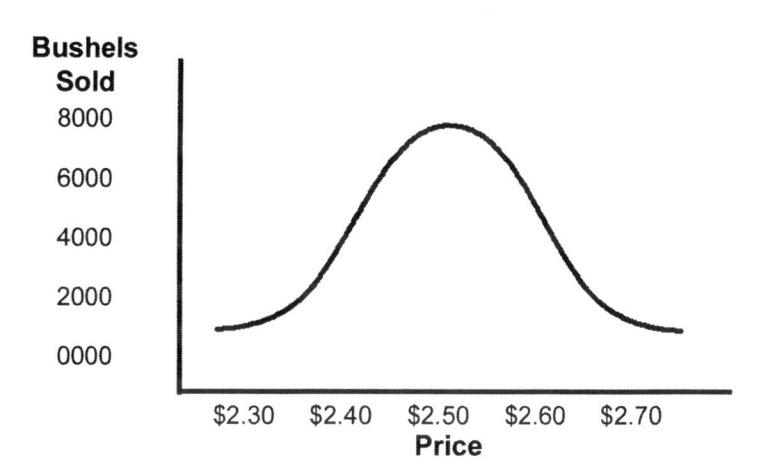

There are many ways to make marketing decisions and often, what is best for you is what fits your personality, your emotional and financial make-up. This method might appeal to you. Its application is easy to apply. More than anything, it may put some discipline in those day-to-day pricing decisions.

You may want to consider including cash sales in your bell curve. Plan a call option buying strategy against at least some of the forward cash sales you make. These calls can help to give you confidence to make the cash sales.

Here are two ways to approach buying these calls: 1) do so at the harvest lows that occur before planting, which is like pre-buying the calls, or 2) buy the calls only if prices start to move to historical highs. For example, in corn, only buy calls if prices get strong enough to rally past $2.90.

Chapter 33
HOW POWERFUL ARE THOSE BIG FUNDS?

Over the years of analyzing commodity markets, I have seen media attention shift from talking about commercial influence in the market to floor traders, then to speculative influence and now to commodity funds. Rarely a day passes without some reference to what the commodity funds are doing and the impact they may have on the markets.

Commodity funds are a powerful force within the futures markets. Over $130 billion is traded in these large futures funds and managed accounts, though only 20% of that managed money is traded using discretion. Most of that money is managed using mechanical systems that strictly react to price movements to generate buy and sell decisions.

In general, managed money moves like a very large ship. It's very hard to turn it or stop it. When the trend of the market is clearly higher, the funds are clearly long. The majority of those funds have no leeway on changing these positions. The bottom line: the majority of FUNDS DON'T THINK!

Many of these large funds lose money and/or have relatively poor performance. Only the top 10% to 20% perform well year-in and year-out. They are the funds that get all the publicity. Funds typically hold well-diversified positions and will offset moderate losses in many markets by participating in big moves in a few markets. Substantial profits only occur in a few markets.

The moral of the story is that you can't believe everything you read. Talking up the funds makes for good copy in news stories. Realize that fund money is NOT necessarily smart money. Marketing decisions should be independent of past, current or anticipated fund actions.

Due to the influence of fund money, price moves are often exaggerated, both to the upside and downside. You can look at this as unwelcome volatility or you can look at the wider price moves as an opportunity to:

- potentially sell at price levels above those justified by supply/demand fundamentals.

- collect greater profits on short hedges and puts.

- increase the opportunities to maximize government farm program payments.

My view is that funds are not going away. Their influence will remain. So look at the opportunity they offer. Their weaknesses and size can be your reward.

New Times, New Rules

Chapter 34
SUCCESS BREEDS SUCCESS: AVOID THE KILLER LOSS

To avoid the "killer loss," it is always important to go with the market price trend, and not fight it. The killer loss is a losing futures trade you hold and keep holding, with the hope that it will turn around. Ultimately, the loss becomes so great that you figure it can't move any further...until it moves against you even more. Finally, you can't take the heat anymore and bail out. This loss kills your equity and your confidence. A killer loss makes it difficult to make good decisions in the future. Typically, people who take killer losses are tempted to take tiny profits on positions because they are afraid their trade will become a loser. Therefore, profitable trades are never held long enough to make gains comparable or greater than the level of the killer loss.

Often I see a hedger or speculator do a good job of trading all year, but they hold just one trade too long and take a killer loss. This trade offsets the profit from other trades or hedges.

With each position you enter, whether it is a hedge or speculative position, you should decide (before you get in) how much risk you're going to take on the position and stick to it. Never change it.

Always remember that success breeds success. Avoid the killer loss.

Section 3
WHERE CAN YOU TURN FOR HELP?

This section talks about the pros and cons of various marketing services and alternatives, and also explores your options of next steps for successful marketing.

WARNING: It contains some self-promotion.

Chapter 35
MARKETING ADVICE PAYS

Yes, as the warning above points out, this chapter focuses on some self-promotion of my business and the business of my competitors. If I didn't believe in what I'm doing, and if I didn't have good competitors, there would be little room for my firm in this marketplace.

My goal in this chapter is to encourage you to seek help with your marketing program. Data shows that most professional farm market advisory firms do a good job for their customers. I am personal friends with the owners and staff of almost every one of my competitors. They are good people who work hard to try to enhance your lives and improve your profitability.

The University of Illinois AgMAS project has produced the most comprehensive track record information on our industry. As with any comprehensive data/research, if you dig deep enough into the data, you will find some things that are not so flattering. However, there is certainly a lot of evidence that using a farm marketing service pays off in the long run.

From 1995 to 2003, the tracked advisory services added $68 per acre in revenue for a farm of half corn and half soybeans. The top 10 advisory firms added $97 per acre in increased revenue over the nine crop years. That's pretty substantial money, and certainly money that most producers would happily accept.

With a total estimated subscription expense over the 9-year period of $1500, a 1500-acre farm would have seen a $150,000 return. That is a 100:1 return.

You should realize that the advice being tracked on these reports is typically advice from daily newsletters and/or published internet reports. It's fairly generic in nature, and written to be understood by the "average" farmer. In other words, these advisory services are producing this kind of added revenue for farmers by using some of the least sophisticated tools, and are trying to write and deliver it in a limited amount of space for producers with a limited amount of knowledge, time and/or patience. It's also a written understanding that there is a general lack of discipline in implementing complicated strategies, so things have to be kept pretty simple. The bottom line is that these advisory services are able to produce these track records while having both hands tied behind their backs.

The services provided are good. Potentially better and more sophisticated advice is available for those willing to: 1) pay to have a personalized marketing consultant or 2) invest the time, effort and, most importantly, the discipline to do it themselves.

Chapter 36
DO YOU WANT PROFESSIONAL HELP?

As we approach the end of the book, I hope you agreed with enough of my opinions and views on marketing that, at this point, you're somewhat open and tolerant to my blatant honesty. Let me tell you why I really wrote this book. I wrote it because I believe farm marketing has changed. I don't believe the simple old methods we all used in the past work anymore. I believe the marketing game has become a very sophisticated game that is more than just a little challenging.

Quite frankly, I believe that very few farmer producers are positioned to win the marketing game without professional help. Very few farmer producers have the time, the desire, or the discipline to consistently develop and implement the sophisticated marketing strategies needed in the years ahead to consistently be a successful marketer. There are some that can market and market well. My purpose in writing this book is to offer an outline to those who are so inclined to undertake this task, an outline to achieve marketing success.

I believe that, in the years ahead, the majority of producers will hire professional farm marketing consultants. This will vary in degree from a simple brokerage relationship, signing over bushels at the local elevator to paying an annual fee to a professional marketing advisor, such as Stewart-Peterson.

You have to be sure that this professional marketing advisor has your interest at heart. They're not only doing your hedging, but they're integrating it and managing it with your cash marketing. They're taking advantage of the carry in the market, and they're watching and managing the basis for you. They have to be aware of your cash flow needs and timing for your cash flow needs. It's complicated, but it can be done. In addition to the above, they need to be using advanced option strategies in combination with futures and cash marketing alternatives. They need to know the forward contracting alternatives available for you at your local elevators, feedlots, co-ops, etc. They have to be willing to recommend those tools when they are to your advantage. They have to customize all of these alternatives and tools to your own personal preferences, your abilities, and your ability to sleep at night.

And above all, they have to communicate with you. The communication has to be regular and understandable. It has to keep you informed, confident and comfortable.

In short, your marketing advisors have to work for you and your interests.

Chapter 37
USE THE RIGHT TOOL

As you move forward in trying to undertake marketing and improve your marketing, it's important for you to select the right tool for the job. For example, a highly skilled carpenter can cut a board square and true with a hand saw, but most people could do much better with a table saw equipped with cutting guides.

The same can be said in farm marketing. Some producers are well equipped to use the "hand tools" and do much of it themselves. For example, some producers are well suited to using a discount broker. If you're a good independent thinker, you're disciplined, you're willing to consistently put in the time to manage your strategies and positions, and you have the willingness and guts to pull the trigger when it's necessary, then you may very well be suited to using a discount brokerage firm. (Author's note: Yes, we do provide discount brokerage.)

At the other extreme, if you're often too busy in the field to pull the trigger and make a decision confidently, or lack the time or desire to develop the ability to use the sophisticated options analysis it takes to economically market your crop, then a more advanced marketing service (such as our Matrix program) may fit. With Matrix, we charge a flat annual fee on a per-bushel basis and low commissions so that all the emphasis is on customizing and managing the marketing tools available in your personal situation.

Between these extremes, we, as well as our competitors, offer many other services, such as newsletters, fee-based brokerage, retail commission-based brokerage and a whole basket of products that your local elevator may be offering. They all have strengths and weaknesses. The key is for you to do the research to learn all you can about these tools and the advisors you might consider. Look at their track records so you can be confident in their performance. And most importantly, look at how well they can provide and meet your individual marketing needs.

Don't just buy from the person calling to sell you something. Often, the most confident and outspoken salesperson will have the worst track record, even though they have the best sales pitch. Buy based on past performance, track record, and a sound approach, not on a bunch of hype or hot air. Look for proof of consistent performance, low volatility and positive returns. The best way to see what someone can do for you is to see what they have done in the past. Look for specific and concrete performance data, not boastful talk.

Chapter 38
BEING A BELIEVER

The first step in accomplishing a task is to believe that it can be accomplished. The old worn-out analogy is the running of a sub-4-minute mile. For many years, no one believed it could be run. Yet, once Roger Bannister did it on May 6, 1954, it quickly became commonplace. It was just a matter of someone believing it could be done, followed, of course, by someone accomplishing the feat.

In my own experience, I find that the biggest challenge in selling our farm marketing advisory services is convincing producers that we can actually do a better job than they could without any help. It's not the competition I worry about, because all of my competitors are pretty good. Rather, my biggest obstacle is apathy. People have tried cash marketing, fundamental analysis, technical analysis, futures, options, average pricing programs at the elevator; they have tried it all. Year-in and year-out, they are still dissatisfied with their marketing. What they've learned over 20 or 30 years of marketing experience is that no amount of experience and no number of commodity brokers or advisors they try ever seems to make enough difference to their bottom line to make it worth the bother.

Look at any marketing seminar these days. It takes three times more effort to get a third as many people to a marketing seminar today as it did 10 years ago. For one reason or another, some people feel that taking time to attend a marketing seminar is a wasted effort. Part of the reason is that most market advisors are basing their presentations on an outlook, which oftentimes proves useless. That's why we use strategy.

Good farm marketing is not a wasted effort. Just weeks before I sat down to write this book, I was giving a speech at Purdue University and, in the six weeks prior to that, corn prices fell so quickly that a typical producer raising 1,000 acres of corn in the central Corn Belt would have lost $150,000 in revenue. Soybean prices fell the equivalent of about $50,000 worth of revenue on 1,000 acres of crop. A 2,000-acre farmer, half corn and half beans, lost $200,000 in revenue in six weeks. Yes, put options looked expensive at the time, but in retrospect, they would have been the greatest buy in the world.

My point: Farm marketing can be done well. Abandon the outlook approach to marketing and embrace the strategy approach, and you will see substantial improvement in your marketing.

Focus on being disciplined in every aspect of your marketing, and you will see substantial improvement. There are professionals that can be hired to assist you. They won't be perfect, but nothing in life is.

Today, any key decision in your farm operation must be made from a sound base of knowledge. If you're looking to hire a professional, thoroughly research their track record, their consistency, their performance and how their style matches yours. Then, be committed to implement the recommendations you're paying them to give you.

To become a good marketer or to become

a good customer of a professional marketing advisor, you have to believe that marketing can be done well, and there is a substantial reward for doing it well. For most producers, past experience would not support this belief. But it is not a misplaced belief, and it is one that is absolutely necessary to the survival and the prosperity of farms across the country.

Chapter 39
SETTING GOALS AND BEING FOCUSED

Throughout this book, I've tried to change your way of thinking and change your perspective in an effort to get you to accept why outlook-based marketing is doomed to failure, and why a strategic approach is so necessary. Making such a dramatic change in your approach to anything can be quite difficult to accept. It's even more difficult to actually integrate and implement. I hope that throughout this book I've instilled in you enough motivation through examples and experiences to paint a picture that will help move you in the direction of change.

I've been in the farm marketing business for over 25 years now, and I'll be in it for many more years. I don't like seeing missed opportunities. Too many farmers are leaving too much revenue on the table and not in their pockets. It's my goal to fix that. It is my goal to make you and other progressive farmers like you a customer of Stewart-Peterson. Not a customer for today or tomorrow, but a customer for life. I would like your kids' kids to be clients of Stewart-Peterson employees' kids' kids someday. If we achieve that, then your farm operation has been a success, and Stewart-Peterson will have been a successful provider of services to you. We can only achieve that success by doing our jobs well and making your job easier.

Stewart-Peterson offers a broad range of products and services to meet producers' marketing needs. If we can be of any assistance to you, we welcome the opportunity.

May all your marketing experiences be consistently profitable.

Appendix A
GLOSSARY

Assignment: Notice to an option writer that an option has been exercised by the option buyer and that the writer is thus required to render the underlying futures contract to the buyer.

At-the-money option: The option whose strike price nearly equals the underlying futures contract.

Automatic exercise: A policy where in-the-money options (options with intrinsic value) are exercised at option expiration.

Basis: The difference between the price of a commodity and the price of a related futures contract.

Basis contract: A cash market contract in which the basis is locked, but the futures price remains open.

Bear spread: To sell a nearby futures or option contract and buy an equal quantity of a more deferred period (e.g., to sell January and buy March soybean futures is a "bear spread").

Bearish: An outlook expecting prices to decline.

Bid: An offer to buy a commodity at a specific price.

Bullish: An outlook expecting prices to increase.

Call option: An option that allows the option buyer the right, but not the obligation, to purchase the underlying futures contract at the specified strike price on or before the expiration date of the option.

Closing range: The range of prices that a futures option contract trades at during the exchange-specified closing price time period.

Delta: The expected change in an option's price given a one-unit change in the price of the underlying futures contract or physical commodity. For example, an option with a delta of 0.5 would change $.01 when the underlying commodity moves $.02.

Glossary

Exercise: The action taken by an option buyer to convert the option to the specified underlying futures contract.

Expiration date: The last day that an option contract trades.

Fence strategy: A strategy that combines calls and puts, establishing a range of possible hedge prices, rather than just one price. You simultaneously buy a put and sell a call option. This creates a "fence" between the put option strike price and the call option strike price.

First notice day: First day on which notices of intention to deliver cash commodities against futures contracts can be presented by sellers and received by buyers through the exchange clearing house.

Flat the market: Occurs when the trader has "closed," "exited," "lifted" or "unwound" any open positions he/she had and now has no positions in the market.

Forward cash contract: A cash grain contract calling for shipment in the future. Some farmers make it a practice to forward contract a portion of their production at planting time.

Fundamental analysis: The use of supply and demand information to analyze and predict price direction and price objectives of a cash, futures, or option market.

Futures position: An agreement to purchase or sell a commodity for delivery in the future: (1) at a price that is determined at initiation of the contract; (2) which obligates each party to the contract to fulfill the contract at the specified price; (3) which is used to assume or shift price risk; and (4) which may be satisfied by delivery or offset.

Hedge-to-arrive contract: Cash market forward contracts in which the futures price has been fixed, but the basis unfixed. Therefore, the final cash price is not fixed until the basis is set at some later point in time.

In-the-money option: An option with intrinsic value; specifically, a call option whose strike price is below the current futures price (or a put option whose strike price is above the current futures price).

Initial margin: The minimum amount of margin required to establish a futures or option position.

Glossary

Last trading day: The day that trading ceases for the nearby (expiring) futures or option contract.

Limit price move: The maximum price move allowed by exchange rules for a specified futures or options contract.

Long: The position created by the purchase of a futures or option contract if there is no offsetting position.

Maintenance margin: The minimum amount of margin deposit required to maintain a futures or option position. (A margin call is initiated if the customer's fund balance falls below this specified amount.)

Margin: An amount of money deposited to ensure fulfillment of a futures or option contract obligation.

Market order: Order to buy or sell futures or option contracts, which is to be executed immediately at the current price trading in the commodity pit.

Offer: An indication of willingness to sell a futures or option at a specific price; opposite of bid.

Offset: Taking a position equal and opposite to the initial transaction to close out a futures or option position.

Open interest: The number of futures or option contracts of a given commodity that have not been offset, delivered against, exercised or expired.

Option: Within the futures industry, a contract that conveys the right, but not the obligation, to buy or sell a futures contract at a specific price for a limited time (see also put, call).

Out-of-the-money option: An option with no intrinsic value; specifically, a call option whose strike price is above the current futures price (or a put option whose strike price is below the current futures price).

Premium: The price paid for an option contract excluding commission transaction fees. The premium is made up of intrinsic value and time value.

Price order: An order in which the customer sets a price limit for the order to be filled at or better than.

Glossary

Put option: An option that allows the option buyer the right, but not the obligation, to sell the underlying futures contract at the specified strike price on or before the expiration of the option.

Range: The difference between the highest and lowest prices that a futures or option contract trades at for a day, week, month, or year.

Ratio spread: Constructed with either puts or calls, the strategy consists of selling a certain amount of near-to-the-money options and buying a larger quantity of out-of-the-money options.

Resistance: In technical analysis, the price area at which prices encounter increased selling pressure or reduced buying interest.

Short: The position created by the sale of a futures or option contract if there is no offsetting position.

Spike: A price move that is sharply above or below the preceding price activity.

Spread: Price difference between two related futures contract months.

Stop order: A buy order placed above the market (or sell order placed below the market) that becomes a market order when the specified price is reached.

Strike price: The price at which the buyer of a call or put may exercise the right to purchase or sell the underlying futures contract.

Support: In technical analysis, the price area at which prices encounter increased buying pressure or reduced selling interest.

Theta: A measure of the rate of change in an option's theoretical value for a one-unit change in time to the option's expiration date. It is a measure of the time decay showing how much an option drops in value as the number of days to option expiration declines.

Tick: The minimum possible price movement, up or down, in a specific futures or options contract.

Volume: The number of purchases or sales of a given futures or option contract made during a specified period of time.

Appendix B
ABOUT THE AUTHOR

Scott Stewart, raised in northern Indiana, received his Bachelor of Science degree in Agricultural Economics from Purdue University. Upon graduation, he became a market analyst for Top Farmers of America, an agricultural publishing and market advisory firm. Mr. Stewart was quickly promoted to the position of Senior Market Analyst, responsible for market strategy and recommendations for publications that reached more than 200,000 farmers. Scott's impressive advisory track record brought him many clients and widespread national recognition.

In 1985, in response to the growing need farmers had for a market advisor who could also help to execute the clients' futures and options transactions, Scott and a business associate started Stewart-Peterson Group. In 2005, the firm celebrated its 20-year anniversary. Over the years, close client relationships and successful products and services helped Stewart-Peterson to achieve an annual growth rate of between 15% and 30%.

Stewart-Peterson has over 25 registered Market Advisors who service both individual and commercial customers through its brokerage and professionally managed marketing programs.

In addition to his commitment to Stewart-Peterson clients, Scott is also actively involved in the agriculture and futures industry. He is Past President and board member of the National Introducing Brokers Association (NIBA), a member of the National Futures Association (NFA), a member of the NFA Hearing Committee, and a member of both the Chicago Mercantile Exchange and the Chicago Board of Trade Ag Advisory Committees.

He has testified on agricultural and futures issues before the Senate Agricultural Committee and, on numerous occasions, before the Commodity Futures Trading Commission.

About The Author

Scott's extensive experience in writing marketing advice, one-on-one consulting, innumerable speaking engagements and countless discussions with savvy farm marketers have heavily influenced his perspective on what works for farm marketers. These life-long experiences are what have led to his "New Rules" on farm marketing.